Practical Mastering

A Guide to Mastering in the Modern Studio

Practical Mastering

A Guide to Mastering in the Modern Studio

Mark Cousins
Russ Hepworth-Sawyer

www.masteringcourse.com

NEW YORK AND LONDON

First published 2013
by Focal Press
70 Blanchard Rd Suite 402, Burlington, MA 01803

Simultaneously published in the UK
by Focal Press
2 Park Square, Milton Park, Abingdon, Oxon OX14 4RN

Focal Press is an imprint of the Taylor & Francis Group, an informa business

Library of Congress Cataloging in Publication Data
Cousins, Mark.
Practical mastering / Mark Cousins, Russ Hepworth-Sawyer.
pages cm
ISBN 978-0-240-52370-5 (pbk.)
1. Mastering (Sound recordings) 2. Digital audio editors.
I. Hepworth-Sawyer, Russ. II. Title.

TK7881.4.C686 2013
781.49--dc23

2012037208

ISBN: 978-0-240-52370-5 (pbk)

ISBN: 978-0-240-52371-2 (ebk)

Typeset in ITC Giovanni Std

By MPS Limited, Chennai, India

Printed and bound in the United States of America by Sheridan Books, Inc. (a Sheridan Group Company).

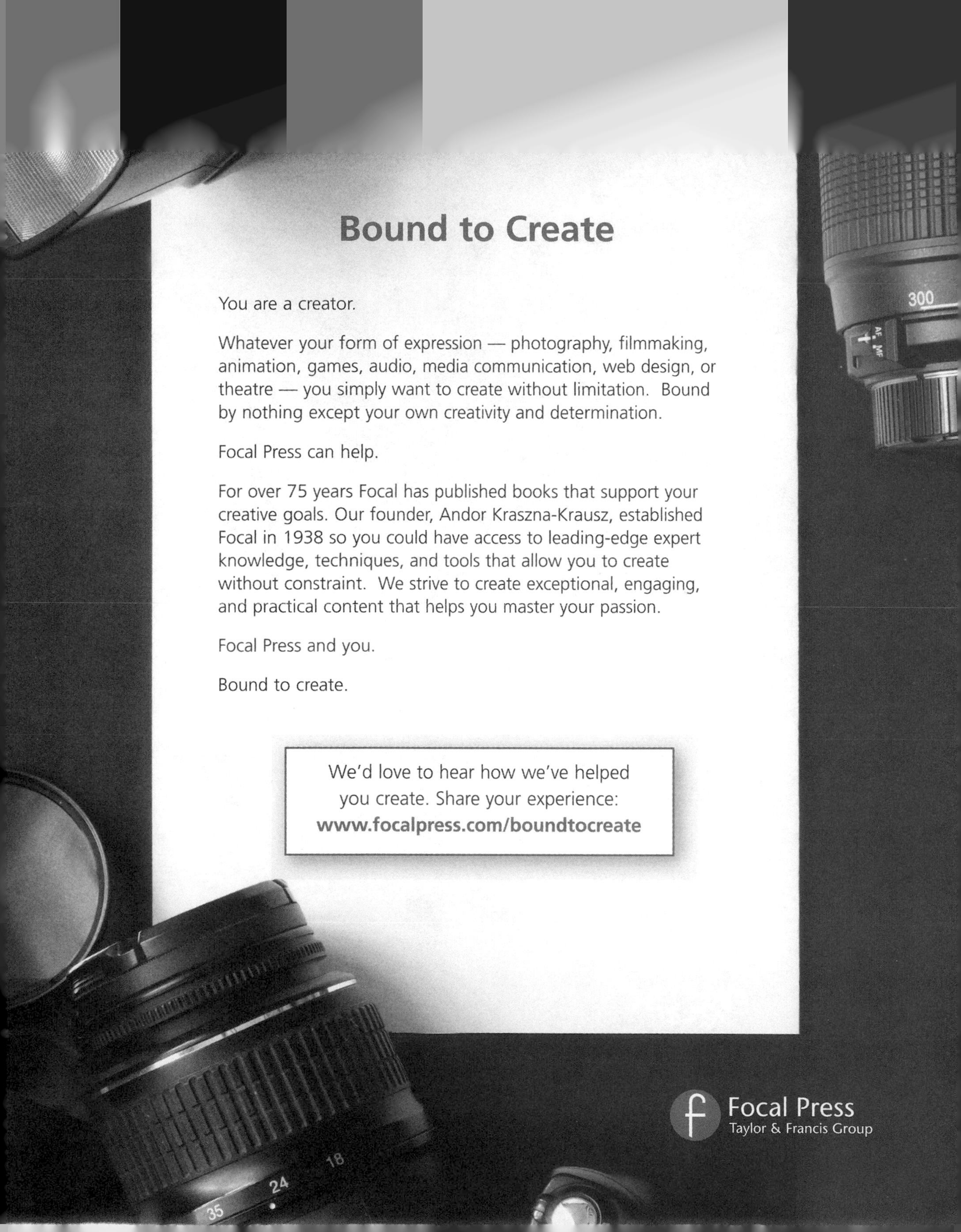

Contents

ABOUT THE AUTHORS xiii
DEDICATION xiv
ACKNOWLEDGMENTS xv

CHAPTER 1 Introduction 1
1.1 An introduction to Mastering 1
1.2 The role of this book 2
1.3 Can I master in [insert your DAW name here]? 3
Knowledgebase 4
1.4 The history and development of mastering 5
1.5 The early days 5
1.6 Cutting and the need for control 6
1.7 Creative mastering 8
1.8 The CD age 8

CHAPTER 2 Mastering Tools 11
2.1 Introduction 12
2.2 The digital audio workstation (DAW) 12
2.3 A/D and D/A converters and the importance of detail 13
2.4 Monitoring 16
2.5 Mastering consoles 17
A – Series of input sources 18
B – Input and output level controls 18
C – Signal processing – including cut filters, M/S and stereo adjustment 18
D – Switchable insert processing 19
E – Monitoring sources 19
2.6 Project studio mastering 19
2.7 Project studio mastering software 20
Apple WaveBurner 21
soundBlade LE 22
Steinberg wavelab 24
2.8 Mastering plug-ins 24
iZotope Ozone 24
Universal audio precision series 26
IK multimedia T-RackS 3 27
Waves L-316 28
2.9 An introduction to M/S processing 29
2.10 M/S tools 30

Using stereo plug-ins in M/S format 30
Dedicated M/S plug-ins 31
2.11 Essential improvements for project studio mastering 33
Acoustic treatment 34
Monitoring 34
Conversion 35

CHAPTER 3 Mastering Objectives 37
3.1 Introductory concepts 38
3.2 Flow 39
3.3 Shaping 40
3.4 Listening 41
Listening interface and quality benchmark 41
Knowledgebase 42
Listening subjectivity versus objectivity 43
Holistic, micro and macro foci in listening 44
Listening switches 45
3.5 Confidence 47
3.6 Making an assessment and planning action 48
3.7 Assessing frequency 48
3.8 Assessing dynamics 49
3.9 Assessing stereo width 49
3.10 Assessing perceived quality 50
3.11 The mastering process 51
3.12 Capture 53
3.13 Processing 53
3.14 Sequencing 54
3.15 Delivery 55

CHAPTER 4 Controlling Dynamics 57
4.1 Introduction 58
4.2 Basics of compression 59
Threshold 59
Ratio 59
Attack and release 60
'Auto' settings 62
Knee 63
Gain makeup 64
Beyond the basics 64
4.3 Types of compressor 65
Optical 65
Variable-MU 66
FET 67
VCA 68

4.4 Compression techniques ... 69
Gentle mastering compression ... 69
Over-easy ... 70
Heavy compression ... 71
Peak slicing ... 72
Glue ... 74
Classical parallel compression ... 74
New York parallel compression ... 76
Multiple stages ... 76
4.5 Using side-chain filtering ... 78
4.6 From broadband to multiband ... 79
4.7 Setting up a multiband compressor ... 80
Step 1: Setting the crossover points and the amount of bands ... 80
Step 2: Apply compression on all bands to see what happens ... 81
Step 3: Controlling bass ... 82
Step 4: Controlling highs ... 83
Step 5: Controlling mids ... 84
Step 6: Gain makeup ... 84
4.8 Compression in the M/S dimension ... 85
Compressing the mid signal ... 86
Compressing the sides of the mix ... 86
4.9 Other dynamic tools – expansion ... 86
Downwards expansion ... 86
Upwards expansion ... 89

CHAPTER 5 Refining Timbre ... 91
5.1 An introduction to timbre ... 92
5.2 Decoding frequency problems ... 94
Broad colours – the balance of LF, MF and HF ... 94
Balance of instruments ... 94
Unwanted resonances ... 94
Technical problems ... 95
5.3 Notions of balance – every action having an equal and opposite reaction ... 95
5.4 Types of equalizer ... 97
Shelving equalization ... 97
Parametric equalization ... 98
Filtering ... 99
Graphic equalization ... 99
Phase-linear equalizers ... 100
Non-symmetrical EQ ... 100
5.5 A Journey through the audio spectrum ... 101
10–60 Hertz – the subsonic region ... 102
60–150 Hz – the 'root notes' of bass ... 103
200–500 Hz – low mids ... 104

500 Hz–1 kHz – mids: tone 105
2–6 khz – Hi mids: bite, definition and the beginning of treble 106
7–12 kHz – treble 107
12–20 kHz – air 108
5.6 Strategies for equalization 109
High-pass filtering 109
Using shelving equalization 111
Understanding the curve of EQ 112
Combined boost and attenuation 114
Controlled mids using parametric EQ 115
Cut narrow, boost wide 117
Fundamental v. second harmonic 118
Removing sibilance 119
5.7 Selective equalization – the left/right and M/S dimension 121
Equalizing the side channel 122
Equalizing the mid channel 124
5.8 Subtle colouration tools 125
Non-linearity 125
The 'Sound' of components 127
Phase shifts 127
Converters 128
5.9 Extreme colour – the multiband compressor 128
5.10 Exciters and enhancers 130
Aphex Aural Exciter 130
Waves Maxx Bass 131

CHAPTER 6 Creating and Managing Loudness 133
6.1 Introduction: A lust for loudness 134
6.2 Dynamic range in the real world 135
6.3 All things equal? The principles of loudness perception 136
6.4 Loudness, duration and transients 138
6.5 Loudness and frequency 140
6.6 The excitement theory 143
6.7 The law of diminishing returns 145
6.8 Practical loudness Part 1: What you can do before mastering 147
Simple, strong productions always master louder 147
Don't kill the dynamics – shape them! 148
Understand the power of mono 148
Control your bass 149
6.9 Practical loudness Part 2: Controlling dynamics before you limit them 149
Classic mastering mistakes No. 1: The 'Over-Limited' chorus 150
Moderating the overall song dynamic 151
Moderating transients with analogue peak limiting 152
Controlling the frequency balance with multiband compression 153

6.10 Practical loudness Part 3: Equalizing for loudness 154
Rolling Off the sub 154
The 'smiling' EQ 155
Controlling excessive mids 156
Harmonic balancing 156
6.11 Practical loudness Part 4: Stereo width and loudness 158
Compress the mid channel 159
Attenuate bass in the side channel 159
6.12 Practical loudness Part 5: The brick-wall limiter 159
How much limiting can I add? 159
Knowledgebase 161
Setting the output levels 162
Multiband limiting 163
6.13 The secret tools of loudness – Inflation, excitement and distortion 163
Sonnox Inflator 165
Slate Digital Virtual Tape Machines 166
Universal Audio Precision Maximizer 167

CHAPTER 7 Controlling Width and Depth 169
7.1 Introduction 169
7.2 Width 170
7.3 Surround or no surround? That is the question 171
Surround-Sound formats and current mastering practices 171
7.4 Stereo width 172
7.5 Phase 172
7.6 M/S manipulation for width 175
7.7 Other techniques for width 176
7.8 Depth 178
7.9 Mastering reverb 179

CHAPTER 8 Crafting a Product 181
8.1 Introduction 181
8.2 Sequencing 182
8.3 Topping and tailing 183
8.4 Gaps between tracks and 'sonic memory' 184
8.5 Fades on tracks 187
8.6 Types of fades 189
8.7 Segues 190
8.8 Level automation 191
Knowledgebase 192
8.9 Other automation and snapshots 193
8.10 Stem mastering 194
8.11 Markers, track iDs and finishing the audio product 196
8.12 Hidden tracks 199

CHAPTER 9 Delivering a Product ... 201
9.1 Introduction ... 201
9.2 Preparation ... 202
9.3 International Standard Recording Code (ISRC) ... 202
Knowledgebase ... 204
9.4 The PQ sheet ... 205
9.5 Barcodes (UPC/Ean) ... 208
9.6 Catalog numbers ... 209
9.7 Other encoded information ... 209
9.8 CD-Text ... 211
9.9 Pre-emphasis ... 212
9.10 SCMS ... 212
9.11 Delivery file formats ... 214
9.12 DDPi and MD5 checksums ... 215
9.13 CDA as master? Duplication v. replication ... 216
9.14 Mastering for MP3 and AAC ... 218
9.15 Delivery ... 220
9.16 Final checks ... 223

CHAPTER 10 DAW Workflow ... 225
10.1 Introduction ... 225
10.2 Method 1: Track-based mastering in a conventional DAW ... 226
Importing audio ... 226
Songs to tracks ... 226
Instantiating mastering plug-ins ... 229
Rendering pre-master files and applying fades ... 230
Final delivery ... 231
10.3 Method 2: Stereo buss mastering in an unconventional DAW ... 232
MClass Equalizer ... 233
MClass Compressor ... 233
MClass Stereo Imager ... 234
MClass Maximizer ... 235
A Word of Caution ... 235

CHAPTER 11 Conclusions ... 237
11.1 The art of mastering ... 237
11.2 Technology as a conduit ... 237
11.3 Head and heart ... 239
11.4 An evolving art form ... 239

INDEX ... 241

About the Authors

Mark Cousins

Mark Cousins works as a composer, programmer and sound engineer (www.cousins-saunders.co.uk), as well as being a long-serving contributor to *Music Tech* magazine. His professional work involves composing music for some of the world's largest production music companies – including Universal Publishing Production Music, among others – and he has had his music placed on major campaigns for brands such as *Strongbow, McDonalds, Stella Artois, Hershey, BT* and *Liptons* as well as being used on hit shows such as *The Apprentice, Top Gear* and *CSI:NY*. He has also had works performed for the Royal Philharmonic Orchestra, the East of England Orchestra, City of Prague Philharmonic Orchestra and the Brighton Festival Chorus.

Mark has been an active contributor to *Music Tech* magazine since issue one. He has been responsible for the majority of cover features, as well as the magazine's regular Logic Pro coverage. As a senior writer, he has also had a strong editorial input on the development of the magazine, helping it become one of the leading brands in its field.

Russ Hepworth-Sawyer

Russ Hepworth-Sawyer is a sound engineer and producer with over two decades' experience of all things audio and is a member of the Association of Professional Recording Services, a Fellow of the Institute For Learning, and a board member of the Music Producer's Guild where he helped form their Mastering Group. Through MOTTOsound (www.mottosound.co.uk), Russ works freelance in the industry as a mastering engineer, writer and consultant. Russ currently lectures part time for York St John University and has taught extensively in Higher Education at institutions including Leeds College of Music, London College of Music and Rose Bruford College. He currently writes for *Music Tech* Magazine, has contributed to *Pro Sound News Europe* and *Sound On Sound*, and has written many titles for Focal Press.

"Dedicated to Joan Margaret Sawyer (1935–2012)."

Acknowledgements

Mark would like to thank...

Thanks to my family – Hannah, Josie and Fred – for their support, patience and amusement over the years, and for keeping me company in the studio from time to time! I am also eternally grateful to my parents for encouraging me to pursue my love of music all those years ago (and for buying me my first synthesizer!). Thanks also to Russ, for his ability to listen to my long ramblings, but most importantly as a long-standing friend – from those early days driving around Sherwood in his Ford Capri, through to the process of writing two books together! I would also like to thank Neil Worley, for taking me on at *Music Tech* and nurturing me though my early days of professional writing, Adam Saunders, for putting up with a fluctuating musical output over several months, and Johnny, for being such a great brother!

Russ would like to thank...

Thanks must first go to my wife Jackie and sons for allowing me to embark on yet another book – thank you for your love and encouragement! Thanks too must go to my parents-in-law Ann and John for all their help and support during the writing. Thanks to Max Wilson for his friendship, honesty and sincere support over the years ... thanks boss! Continual thanks owed to: Iain Hodge at London College of Music; Craig Golding at Leeds College of Music; Ben Burrows and Rob Wilsmore at York St John University.

Thanks must be mentioned to the following for their friendship and guidance in the world of audio and mastering (they have not contributed directly to this book, but indirectly from years of conversations over cups of tea and beers!). Sincere thanks must be provided to the mastering friends in the industry namely: John Blamire, Dave Aston, Ray Staff, Paul Baily, Bob Katz plus all the mix engineers and clients I've had the pleasure to work with. Specific thanks are extended to Barkley McKay, Rob Orton, Tom Bailey, Adrian Breakspear, Iain Hodge, Craig Golding and so many others for helping me think about the process over many years. Also a quick shout out to Catherine 'Parsonage' Tackley, Tony Whyton and Ray, Sue, Toto and Billy Sawyer.

Finally I'd like to thank Mark for his long-standing friendship and support. It was a long time ago when we started all this music stuff professionally hey?

CHAPTER 1

Introduction

In this chapter

1.1 An introduction to Mastering 1
1.2 The role of this book 2
1.3 Can I master in [insert your DAW name here]? 3
Knowledgebase 4
1.4 The history and development of mastering 5
1.5 The early days 5
1.6 Cutting and the need for control 6
1.7 Creative mastering 8
1.8 The CD age 8

1.1 An introduction to Mastering

Compared to all other parts of the production process, mastering is surrounded by the greatest amount of miscomprehension and mystery. 'What makes a finished master sound like a "record"?' 'How do professional mastering engineers create such loud masters?' and 'Can mastering really turn my lackluster demo into something listenable?' – the list of questions about possibilities and skills of mastering are almost endless. In truth, few of us can ignore the important role mastering plays – whether you're demoing songs in a home studio, or working as a professional musician/engineer – so it's well worth separating fact from fiction.

In its broadest sense, mastering forms the final part of the production process – the last opportunity to finesse the sound of a recording and the point at which we deliver a product ready to be replicated and delivered to the consumer. Over time, and with the rapid changes in technology over the last 40 years, the exact role and function of mastering has changed. Where mastering engineers might have once dealt with a cutting lathe and a vintage valve compressor, today's engineer might be delivering internet-ready content mastered through a range of cutting-edge plug-ins. Either way, mastering is most effective when the music is conveyed as effectively as possible.

Arguably the most important development in mastering is that all musicians and engineers, to a lesser or greater extent, have to actively engage in the mastering process. The record industry of old – where just a handful of mastering engineers acted as gatekeepers to the outside world – has long gone, and instead, musicians and recording engineers have any number of avenues through which their music can be released. With budgets being stretched, release dates looming and 'off-the-shelf' recording technology delivering high-quality results, it's a logical conclusion that the mastering process slowly becomes part of everyone's recording workflow, just like the tasks of tracking or mixing.

Of course, there's still a role for a professional mastering engineer – principally as a means of bringing some 'fresh ears' to a project, but also in utilizing some well-honed experience in the final stages of the production process. There's also a wealth of specialist equipment that they can bring into the equation, most of which isn't appropriate for conventional music production activities, or is simply too expensive for most of us to afford. That said, few people can choose to ignore the role mastering plays, and everyone can benefit from a greater understanding of the contribution that mastering makes to the record-making process.

1.2 The role of this book

Given a need to 'self master', this book is designed to equip you with all the skills relevant to working in today's music industry. It recognizes the fact that a majority of engineers choose to mix and master 'in-the-box', but that there's still a role for specialized mastering hardware, and indeed, a dedicated mastering engineer where appropriate. Most importantly, it illustrates what all of us can realistically expect to achieve with access to a reasonable set of production tools (in other words, a DAW and a small, well-chosen collection of plug-ins), as well as defining the limitations and potential pitfalls that such an approach can deliver.

If you're starting out in music production, therefore, this book is a ideal way of understanding how mastering can improve your music and how you can best integrate your output with the rest of the audio world. Starting from the basics, we'll look at the key processes and techniques behind mastering – whether you're using a compressor to better control the dynamics of a piece of music, or assembling a finished DDP (Disc Description Protocol) master ready for replication. We'll also guide you through the key issues of the day, including the ever-present loudness war and what this means to audio quality and music in general.

For the more experienced user, the book illustrates a more refined understanding of the tools you have at your disposal – understanding how *equalization*, *compression* and *limiting* interact to transform a basic mix into a finished master. Rather than just skipping through presets, you'll gain better understanding about the role and relevance of each individual parameter, and how you can tweak these controls in a meaningful way to achieve a defined objective. You'll also form a better understanding of the wider principles of mastering – from the ethos and thinking behind 'professional quality' audio, through to the technical details, like delivery formats and so on, that are so important to how you interact with the wider audio industry.

1.3 Can I master in [insert your DAW name here]?

In writing this book we've tried to strike a balance between universal information (in other words, good audio mastering practice that can be applied in almost any *digital audio workstation*, or DAW), and advice and guidance that relates to specific applications and plug-ins. The majority of the book deals with what we consider to be the key skills of mastering – focussing on areas such as dynamics and timbre as well the mechanics of putting together a finished product. Although the precise nature of what you can achieve will vary from DAW to DAW, it's interesting to note the universality of many of these skills. At the end of the day, all DAWs offer some way of editing, equalizing and compressing an audio file, which means that you can at least start to practice and apply some of the principles of mastering.

Knowledgebase

Mastering: a step-by-step guide

The first stage of mastering is generally concerned with the sound of the finished master, using tools such as compression, equalization and limiting to create a balanced and homogenous listening experience. Deficiencies in the mix (such as a weak bass or an over-aggressive snare drum) can often be refined by the considered application of signal processing, and a good engineer can also ensure that the track has a suitable dynamic range and timbral balance so that it translates across a range of listening environments. It's also important that the album as a whole sounds balanced – that tracks have an even loudness, for example, and that they share a similar tonal colour.

Once the sound of the masters is established, an engineer can look towards editing and ordering the final playlist for the tracks. The beginnings and ends of tracks will often need some attention – removing unwanted noise at the track's start, for example, or applying a fade-out over the last chorus. Assuming you're mastering an audio CD, it's also important to consider the gaps between tracks and additional factors such as ID markers that might help the listener navigate the music that's presented to them.

The final stage is delivery: ensuring that the final product is suitable for the next stage of production. If you're creating a CD, for example, you might need to deliver the master in DDP format (Disc Description Protocol), or you might simply be delivering audio files (in a given format) if the material is

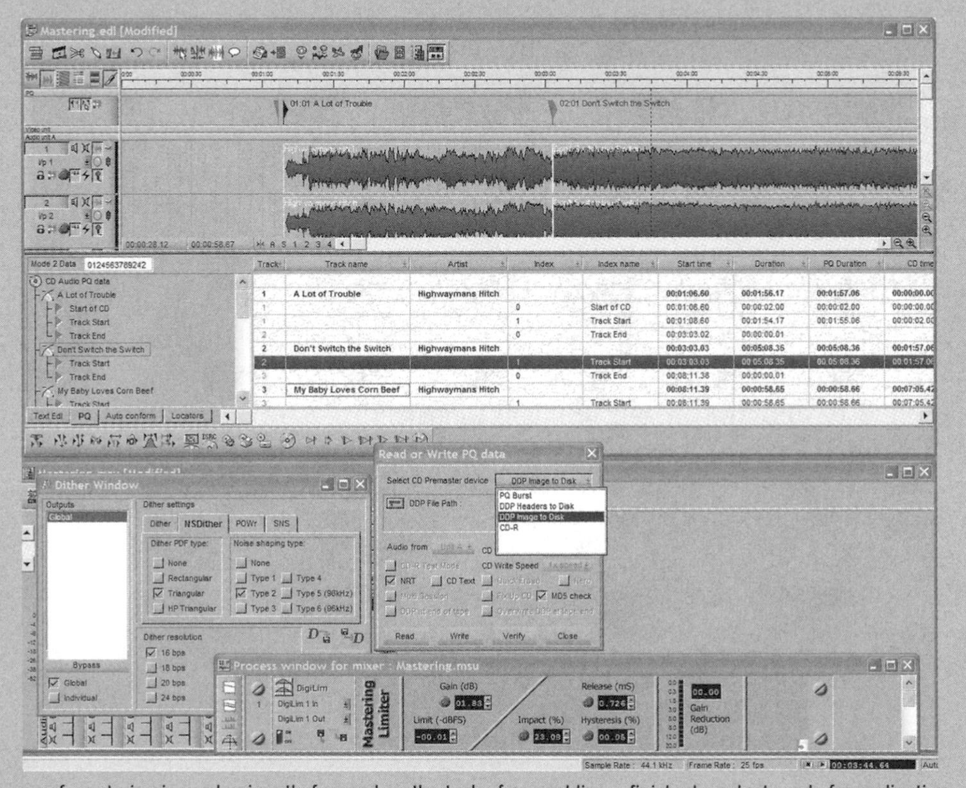

The process of mastering is predominantly focused on the task of assembling a finished product ready for replication.

going to be used on the Internet. Either way, it might also be important to embed additional metadata – such as ISRC codes, composers' names and so on – so that your music can be tracked and monitored correctly in the digital universe.

Towards the end of the book, though, we've also included a chapter that looks at a number of specific applications – highlighting the key operational points you need to be aware of, as well as the principle differences that exist between the different working methodologies. To keep the information succinct it isn't an exhaustive explanation, but instead offers an illustration of how the various techniques we've described can be applied in a number of leading DAWs and plug-ins. Given the flexibility inherent in most platforms, and the growing number of new plug-ins released on a daily basis, it's well worth you exploring beyond this and seeing just how much you can achieve!

1.4 The history and development of mastering

Arguably the best way of understanding mastering today is to look at how the role of mastering has changed and evolved over the years. Accepted practices that we assume to be *de rigueur* have developed over many years to become the art form we now appreciate. Of course, to adequately provide a proper and detailed history of mastering would necessitate a book in its own right, so hopefully this provides an insight into the changes that have happened rather than being an exhaustive exploration.

Mastering as we know it today has been borne out of a number of technological and aesthetic developments which, to the untrained eye, seem to be far removed from the origins of the role. Yet over time these developments in technology, distribution and listening habits have resulted in a set of changes in the world of mastering. We now pause to take a concise look at the history of mastering, hoping to bring context to an already misunderstood art form.

1.5 The early days

Back in the earliest days of recording, a sound engineer principally worked as a 'jack-of-all-trades', being both the 'recording engineer' and the 'cutting engineer' as part of the same process. Equipment was limited, with often just a single microphone to capture a full ensemble, which was then used as the sound source to cut the performance direct-to-disc. The recordings were imbued with energy and plenty of imperfections, but engineers still tried their best to ensure that they captured the true spirit of the original lineup and performance.

A significant shift came with the invention of magnetic tape, which meant that the recording process could be separated into two distinct stages – that of capturing the performance, and secondly, the process of cutting the performance to disc. In effect, what we now know of as a mastering engineer started life as a transfer engineer, ensuring that the music recorded to tape was transferred to a subsequent format as effectively as possible. Their concerns weren't necessarily artistic, but instead were to ensure the transfer retained the sonic qualities of the original source recording as effectively as possible. Even at this stage, you can see why high-quality monitoring is so important to a transfer engineer (latterly to become a mastering engineer) as their job relied on a near-flawless transition!

The revolution offered by magnetic tape culminated in the development of multi-track recording, which led to a further demarcation of the recording process – this time creating a distinct third role of a mixing engineer. Over the 1970s and 1980s, therefore, the art of the sound engineer evolved in completely distinct and separate ways, with recording engineers concentrating on capture, mixing engineers concentrating on balance and mastering engineers becoming highly skilled at the process of cutting music to vinyl. Technical standards were understandably high and engineers honed their skills in a defined part of the production process, arguably creating a real 'golden age' of recording.

1.6 Cutting and the need for control

The skillset that evolved in line with the procedures and practices of vinyl cutting were born both from technical limitations of the format, and, latterly, the desire to exploit the creative potential of the mastering process (which we'll return to later on). To ensure these limitations were avoided, cutting engineers would be responsible for transferring the master tape to an acetate without it causing issues with later playback. For example, too much bass could cause many

What is loudness?

One of the biggest differences between a finished mix and a commercial release is its perceived loudness. Put simply, there's a big difference between the electrical measurement of a signal's amplitude (the peak level a mix might meter at, in other words) and our perception of its volume, otherwise known as loudness. For example, two different audio files might both peak at 0dBFS, yet our ears will often perceive them as being radically different from each other in respect to loudness.

Given that a mix's loudness is directly linked to how excited we feel about a piece of music, it's no surprise to find that mastering engineers actively manipulate loudness to make a master sound 'better' to the listener's ears. By reducing short transient peaks, limiters can raise the average level of a master and thereby fool our ears into thinking that they're hearing 'louder' music. The harder the limiter is pushed, the greater the perceived loudness becomes.

However, like all good things, there's a cost to the manipulation of loudness. First and foremost, restraint often seems to get pushed out of the equation, with many artists and record companies aggressively limiting masters so that they appear to be louder than the competition. As the limiter works harder, audio quality quickly starts to be eroded, restricting the dynamic range, reducing transient detail, and even worse, adding distortion. In truth, therefore, loudness is a phenomenon that needs to be exploited in a balanced and considered way.

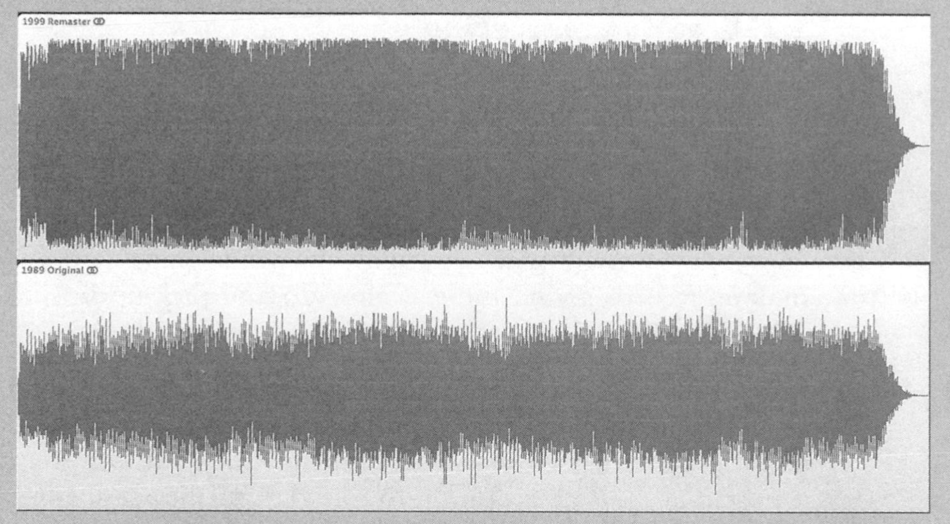

An increasing demand for loudness has meant the sound of recorded music has changed over the last 20 years.

styli to leave the comfort of the correct groove, and cutting engineers might use high-quality filters to soften the bass load on a track, or at a specific point, to ensure that a decent level could still be achieved without causing a jump.

Interestingly, these limitations have somewhat embedded themselves into what we commonly accept as 'good practice' within audio production, arguably informing processing such as recording and mixing as well as mastering. With the fall of vinyl sales, though, the process of cutting records is now the truly specialist end of the market, and the proof is that there are fewer than 100 lathes in the UK working today. The problem is that the coveted cutting machines – mainly the Neumann VMS series – have not been made for many decades, and as far as we know there's only one person in the UK able to keep them going.

The concept behind the lathe is to cut a record from a lacquer disc from which a pressing master is made. For this medium, the mastering engineer (or cutting engineer to use the older term) will likely make some additional changes to the audio to ensure that it works best on the vinyl medium. If you're after vinyl mastering or duplication then you will need to liaise with someone who has access to a cutting lathe, and have a master cut by an experienced mastering engineer.

1.7 Creative mastering

The introduction of much needed processing, such as simple EQ, began a slow trend towards the transfer engineer moving from repairing issues with the material for the medium, to slowly fulfilling a creative role in the process of making records. Similarly, the term changed from the simple 'transfer engineer' to the 'mastering engineer'. Over time, these engineers would gain a reputation for making records sound the way they do and pick up business based on recommendations or from 'fans' of records cut before.

Concurrently, new techniques and possibilities were being developed based on new equipment that would enhance the audio coming from the studio. The idea that the cutting engineer was a potential *meddler* regarding the quality of the studio mix was set aside as people began to appreciate the improvements and repairs that could be made in the last step of the creative process.

1.8 The CD age

By the early 1980s, at the dawn of the CD age, mastering had a new set of limitations and sonic challenges to overcome. Early CD players were sonically cruder than now (conversion technology has improved over the years), so once again engineers would need to ensure the audio was 'ready for the medium'.

Ironically, digital audio was lauded as the final solution to the hiss that we were commonly used to with cassettes, and to stop the artefacts common to the humble turntable. However, digital audio presented a whole new range of issues, such as quantization errors at quiet levels and a fixed top level. In analogue recording we'd become used to keeping levels as hot as would be sensible to avoid the inherent noise floor. The odd peak was not detrimental as harmonic distortion would occur and become a welcome part of many recordings. Digital on the other hand presented us with a very fixed upper limit of 0dBu and exceeding this had disastrous sonic qualities.

This should have led to slightly new techniques being adopted and a new management of dynamic limitations. It could be said that the introduction of this maximum limit began 'loudness war 2' (the first war was the permanent challenge of making a louder record without a maximum limit).

A new set of skills in preparing masters intended for CD using U-Matic tape and Sony 1610 (later 1630) converters developed and remained the mainstay of transferring to a glass master. Eventually mastering was done using digital audio workstations such as Sonic Solutions, later ditching the 1630 for DDP (Doug Carson & Associates' *Disc Description Protocol*) sent to the pressing plant on exabyte tape.

For many years CD Mastering has remained largely based around the DAW, with constant development and flavours of software, all of which we will introduce later in this book. New practices and ways of working have developed in tandem with the complex and fascinating world of mastering we now experience.

For further information, please consult the accompanying website: www.masteringcourse.com

Pre-mastering

What we engage with in the mastering studio today is more typically what many should perhaps call pre-mastering. Technically 'mastering' as a term relates to the physical creation of the replicated copy intended for sale in the shops. Hence the activity of preparing the audio for this medium is more accurately pre-mastering. Commonly pre-mastering is only ever referred to as mastering.

CHAPTER 2

Mastering Tools

In this chapter

2.1 Introduction 12
2.2 The digital audio workstation (DAW) 12
2.3 A/D and D/A converters and the importance of detail 13
2.4 Monitoring 16
2.5 Mastering consoles 17
A – Series of input sources 18
B – Input and output level controls 18
C – Signal processing – including cut filters, M/S and stereo adjustment 18
D – Switchable insert processing 19
E – Monitoring sources 19
2.6 Project studio mastering 19
2.7 Project studio mastering software 20
Apple WaveBurner 21
soundBlade LE 22
Steinberg wavelab 24
2.8 Mastering plug-ins 24
iZotope Ozone 24
Universal audio precision series 26
IK multimedia T-RackS 3 27
Waves L-316 28
2.9 An introduction to M/S processing 29
2.10 M/S tools 30
Using stereo plug-ins in M/S format 30
Dedicated M/S plug-ins 31
2.11 Essential improvements for project studio mastering 33
Acoustic treatment 34
Monitoring 34
Conversion 35

2.1 Introduction

The tools of any profession – whether a joiner or a brain surgeon – are vital to a successful outcome, and mastering is certainly no exception. If you're already involved in music production, many of the tools used for mastering – including compressors, equalizers and digital audio workstations (DAWs) – should be immediately familiar, although given the outcomes of mastering there are often some subtle but important differences to be noted.

For the purpose of clarity, we're going to look at two contrasting approaches. Firstly, we'll explore the key components of a professional mastering facility, looking at the specific choices of equipment, and the reasons and benefits offered by these working approaches. On a more pragmatic note, we'll also take a look at a growing number of options available in the project studio: from flexible music production software that can also handle mastering activities, through to a growing number of mastering-grade plug-ins that offer near-identical results to dedicated mastering signal processors.

Understanding the tools of mastering is important for several reasons. First and foremost, these tools form the principle components of our given workflow – understanding where an audio signal might begin its journey, the devices it might flow through and where its eventual destination resides. If you're working in a music studio, it's also important to be aware of where your limitations might lie – to understand what's achievable with the equipment at your disposal, but to also be wary of making certain key decisions. Of course, should you decide to augment your existing studio to make it more 'mastering-friendly' you'll also have some understanding of where your money is best spent.

2.2 The digital audio workstation (DAW)

Just like any recording studio, the hub of any mastering facility will be its DAW. However, rather than using a traditional music production system such as Pro Tools or Logic Pro, the majority of professional facilities will use a dedicated mastering workstation such as Sonic Solutions, Sadie or Pyramix. In many ways, these mastering workstations contain similar features to those of the conventional DAW, including the ability to edit and compile audio files, as well as the option to apply effects such as EQ or compression.

In addition to these principle audio-editing features, a mastering workstation will also include a range of tools specifically related to the process of assembling a finished Red Book CD. From the ability to enter ISRC codes (International Standard Recording Codes that signify various meta data in relation to the CD), through to the ability to deliver the final product in the form of a Disc Description Protocol export (DDP), mastering workstations are built with one final product in mind – the audio CD. It's also interesting to note the workstation's abilities in conventional audio tasks, often demonstrating particular prowess and flexibility in areas such as crossfades and fade editing where a mastering engineer will have some exacting requirements.

2.3 A/D and D/A converters and the importance of detail

The process of conversion – taking the signal in and out of the digital domain – is vital for any part of the production process, although the issues of quality come into sharp focus during mastering. Most of us will have encountered converters as part an audio interface connected to our DAW. A/D (or analogue to digital) converters will take the analogue signal being fed into the inputs of our audio interface and convert it into the stream of digital 'zeros and ones' that our DAW works with. Likewise, the D/A converter on the output of the interface converts the digital audio signal into an analogue signal that our monitors can amplify.

When it comes to recording and mixing, the quality of conversion is important to the end results, although arguably not as vital as it is in mastering. For example, assuming that the recording is to be mixed 'in-the-box', it's only the A/D conversion on the input path that will be embedded into the recording. In this case the output D/A is only used for monitoring, so as long as you're hearing a reasonable representation of the musical balance, you're in relatively safe hands.

So why is good quality A/D and D/A conversion so important to mastering? Well, this isn't the first or last time we need consider the role of subtlety and detail in mastering. To master correctly, we need to hear every last detail of a recording in all its naked glory: the detailed transients of a drum kit, the shimmer of an acoustic guitar, the precise release of compressor, the exact colour of a reverb and

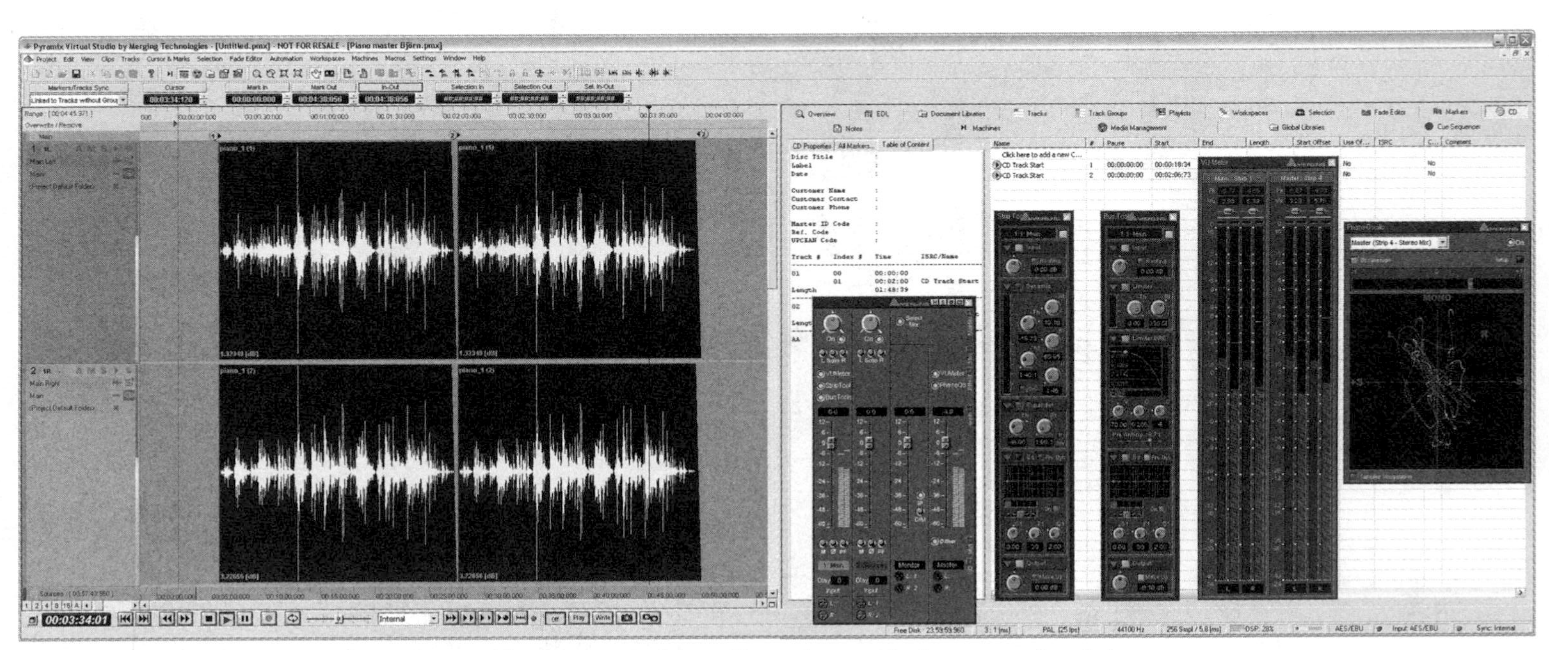

A dedicated mastering audio workstation tends to provide a more appropriate working environment for the process of mastering.

The quality of A/D and D/A conversion comes into sharp focus when mastering.

so on. Good decisions in mastering are ultimately informed by your hearing, and even the smallest distortion delivered by poor-quality conversion can cloud your judgment in some rather extreme ways.

Remember, unlike the process of mixing, whereby you might carry hundreds of modifications to a variety of audio signals, mastering often involves a small number of well-placed modifications – +0.5 dB boost at 12 kHz, for example, or 2 dB of gain reduction on a compressor. Put simply, if you can't hear the subtleties at this level of detail, you can't expect to deliver a professional-sounding master. While this might be partially explained by a lack of experience, it's also interesting to note how several components found in a traditional music studio might impede this – from the quality of the D/A conversion, through to areas such as monitoring and acoustic treatment that we'll deal with in more detail as we move through this chapter.

As well as the quality of conversion in respect to monitoring, we'll also need to consider its direct role and impact on the audio quality of the final master. Unless the mastering is achieved entirely in the digital domain, the sound of the converters will inevitably end up in the sound of the finished product. Most mastering engineers put a great deal of emphasis on the positive contributions that analogue signal processing can impart on a piece of music, so it's important that the conversion used to take a signal out to the analogue domain and back in again retains all the integrity of the original recording.

Compression and limiting

A *compressor* is a form of automated gain control device – in other words, a signal processor that modifies the volume of a signal in response to the input levels. A compressor can restrict and control the dynamic range of a master – turning down 'loud' signals that exceed a stated threshold (measured in dB), which in turn allows you to raise the average level of signals beneath the threshold (quiet signals, in other words).

In mastering, three different types of compressor are used, each of which modifies the dynamics in a slightly different way. A single-band, or *broadband*, compressor works with the entirety of the mix, often used to provide subtle dynamic control and a sense of 'glue'. A *multiband* compressor, on the other hand, splits the mix into two to four frequency bands and applies compression to each band separately before summing all the bands to form an output. Applied in this way, a multiband compressor can apply different compression to distinct parts of the frequency spectrum – applying more to the low-end, for example, while applying a lighter touch in the highs.

The final type of compressor is a *limiter*, which is an aggressive type of compressor that is designed to respond to peak signals. Pushed hard, a limiter can add distortion, but it's also a vital tool for increasing the subjective *loudness* of a master.

Compressors and limiters form an essential part of the mastering process and are used to control and refine the dynamics of the master.

2.4 Monitoring

Of course, the greatest D/A converters in the world won't be able to reveal any hidden detail unless they're connected to an effective pair of full-range monitors. As you'd expect, good monitoring is the crucial lynchpin in any mastering environment, with a large part of the studio's budget tied up in the main pair of monitors. Interestingly, there are also some distinct differences between the monitoring practices found in a dedicated mastering facility and a music studio.

The principle difference is that mastering engineers have a preference for mid-field and far-field monitors, as opposed to the near-field monitors commonly found in production studios. To a large extent, near-fields are a form of pragmatic compromise, giving an engineer (or music listener) an acceptable bandwidth and enough power to keep them happy. The two main areas that suffer are the bass response, with near-fields often being severely compromised beneath 60 Hz or so, as well as the frequency response around the crossover, which is used to divide the audio signal between the woofer and tweeter.

Mid-field monitors, however, are much closer to a no-compromise performance, although their price reflects this accordingly. They typically use three drivers, negating the nasty crossover issues that occur around 2–3 kHz, which can often impinge on the presentation of key elements such as vocals. There's also plenty of power behind them, allowing mid-fields to be placed further back in the control room, widening the sweet spot and giving the monitors plenty of headroom before they go anywhere near distortion. However, because mid-fields are much further away from the engineer, the room will have more impact on the sound, making it essential that acoustic treatment is in place.

Three-way mid-field monitors tend to be the preferred option for monitoring in mastering applications.

2.5 Mastering consoles

From the perspective of mastering, the needs and objectives of a console are radically different to that of mixing. When you're mixing, of course, channel count is everything, so that you can divide all the instrumentation onto separate channels and balance them accordingly. In mastering however, the need for multiple channels is largely pointless, and instead what you want is finessed control over a few simple features – whether it's switching 'in' and 'out' inserts for example, moving between a various monitor sources, or finely tuning the stereo width.

As you'd expect, there are some big differences between console designs, but here are the principle features you'd expect to find on a dedicated mastering console.

A – Series of input sources

The input to the console can come from a variety of playback sources, including DAW, DAT, reel-to-reel tape and so on. The input source is what the console will process – both by features on the desk and insert processing connected to the console.

B – Input and output level controls

Trim controls on the inputs and output of the console allow you to drive the signal processing and subsequent D/A conversion appropriately. It might be that a compressor needs to be driven harder, for example, or the output levels slightly reduced to negate distortion on the converters.

C – Signal processing – including cut filters, M/S and stereo adjustment

The console may include some form of rudimentary signal processing, usual addressing basic 'housework' tasks such as high- and low-pass filtering as well as adjustments to the stereo width. A large number of dedicated mastering consoles also feature M/S matrixing, allowing discrete controls over the 'middle' and 'side' components of a stereo signal (for more information on M/S see Section 2.9 later in this chapter).

D – Switchable insert processing

A series of pre-assigned inserts can be switched in and out of the signal path, allowing you to quickly audition their individual and collaborative contributions to the output. Inserts will usually be in the form of analogue compressors, equalizers and limiters. If the console has M/S matrixing, it might be that the insert processing can be patched into the M/S path.

E – Monitoring sources

The console will allow you to switch between a variety of monitoring sources, allowing you to compare processing that you've applied along the way. A trim control provides a means of attenuating certain monitor sources, which is often essential to avoid fooling yourself as to the the perceived benefits of compression and equalization that are actually just increasing the overall signal levels.

2.6 Project studio mastering

In an ideal world it would be desirable to have access to all the equipment held in a dedicated mastering facility. However, for most music production facilities it's neither cost effective nor practical to have a full complement of analogue 'mastering grade' audio tools. With these limitations in mind, how do you go about the task of mastering using the tools a typical music studio has access to? Are there additional applications and hardware that you need to consider, and what are the limitations inherent in your choices in hardware and software?

What follows is a pragmatic guide to equipping yourself for the task of mastering, identifying both the tools that you can't afford to be without, but also ways in which you could seek to improve the quality of your mastering with a modicum of long-term investment. Of course, it would be unrealistic to think that with just a few hundred pounds you can transform you spare bedroom into a dedicated mastering facility, but with a little considered investment you can at least produce an output that is proportionately closer to a professional release.

2.7 Project studio mastering software

Most off-the-shelf DAWs should allow you to carry out a large part of the mastering workflow in respect to signal processing and editing. Indeed, if you're mainly interested in the 'sound' of mastering, rather than assembling a finished production master ready for replication, there's a lot to be said for staying within the domain of your usual working environment. Applications such as Pro-Tools, Logic and Cubase can all be re-appropriated for mastering tasks, although some of the more unusual production solutions – such as Ableton Live or Reason – aren't so appropriate. Even so, you'll still be able to insatiate some degree of stereo buss processing, potentially allowing you to export 'mastering grade' audio files.

Equalizers

An equalizer is a signal processor that is used to modify the timbre of a master, boosting or attenuating different parts of the frequency spectrum. Traditionally, most people's first encounter with EQ is in the form of a treble and bass control found on a typical hi-fi. Technically speaking, these treble and bass controls are a form of *shelving equalizer*, used to modify the broad timbral qualities of an input signal, rather than providing any detailed control. Parametric equalization – where the user has control over the exact frequency, the amount of boost or cut, and the width of the modification – is a more precise tool that lets you modify detailed parts of the spectrum.

Although it's possible to use any equalizer for the purposes of mastering, dedicated mastering equalizers (either in hardware or software form) tend to deliver the best results for a number of reasons. Firstly, the interaction and curve of the various frequency bands are specifically engineered for the task of processing entire mixes rather than individual instruments – in short, the EQ is empathetic to a finished track rather than its parts. Secondly, you'll often find discrete controls for both the left- and right-hand channels, allowing you differentiate between the equalization applied to either side of the stereo image. Finally, hardware equalizers often have stepped controls, allowing the engineer to precisely recall settings from one session to the next.

Equalizers allow you to shape and refine the timbral qualities of your finished master.

Given the ability to edit audio files and apply signal processing, most studio-based DAWs are suitable for basic mastering activities.

Where most off-the-shelf DAWS fall down, however, is when it comes to creating a finished Red Book master CD, or, going the next stage up, a proper DDP master used by professional CD replication facilities. Here though, you may find it beneficial to look at a number of different applications designed to bridge this divide. While they'll struggle to complete with the likes of Sadie or Pyramix, you may find them as useful tools in expanding your existing setup without having to completely change your studio.

Apple WaveBurner

WaveBurner is one of the suite of applications designed to accompany Logic Pro, essentially providing a full set of tools to create a Red Book master from audio mixes exported from Logic. Once the audio files have been imported into WaveBurner, you can apply a series of audio processing tools – including many of Logic's internal plug-ins and any third-party audio units – all without having to re-render your master. As well as creating finished Red Book audio CDs,

WaveBurner will now also export using the DDP protocol, making it a surprisingly flexible solution.

soundBlade LE

soundBlade LE's heritage stretches back to Sonic Solutions, which was one of the earliest DAWs initially developed by former Lucasfilm employees. The system is one of the most well-respected two-track editors around, with a host of export options including the all-important DDP protocol. As your needs grow, you can also consider upgrading to soundBlade SE and soundBlade HD, which increase the number of available tracks (making soundBlade suitable for stem mastering, or 5.1 work) as well as an expanded set of export options.

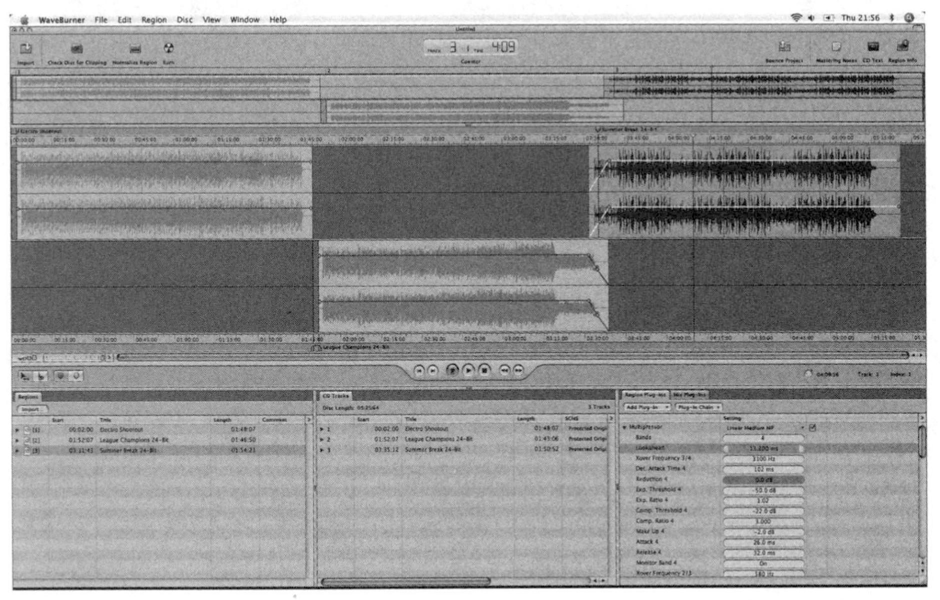

WaveBurner is bundled with Logic Pro and is designed as means of creating Red Book masters from mixes exported from Logic.

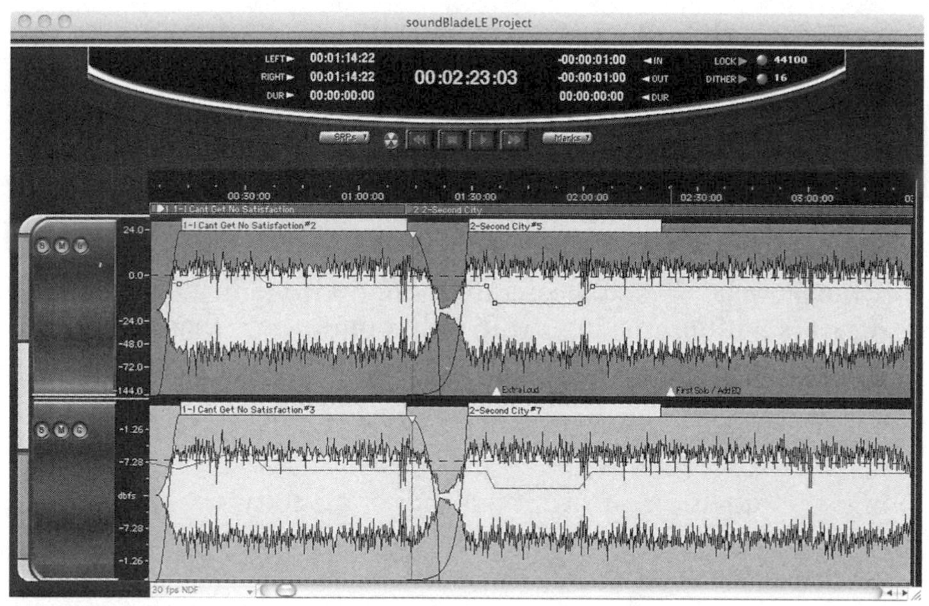

SoundBlade LE is an entry-level solution that has evolved from a dedicated mastering platform - Sonic Solutions.

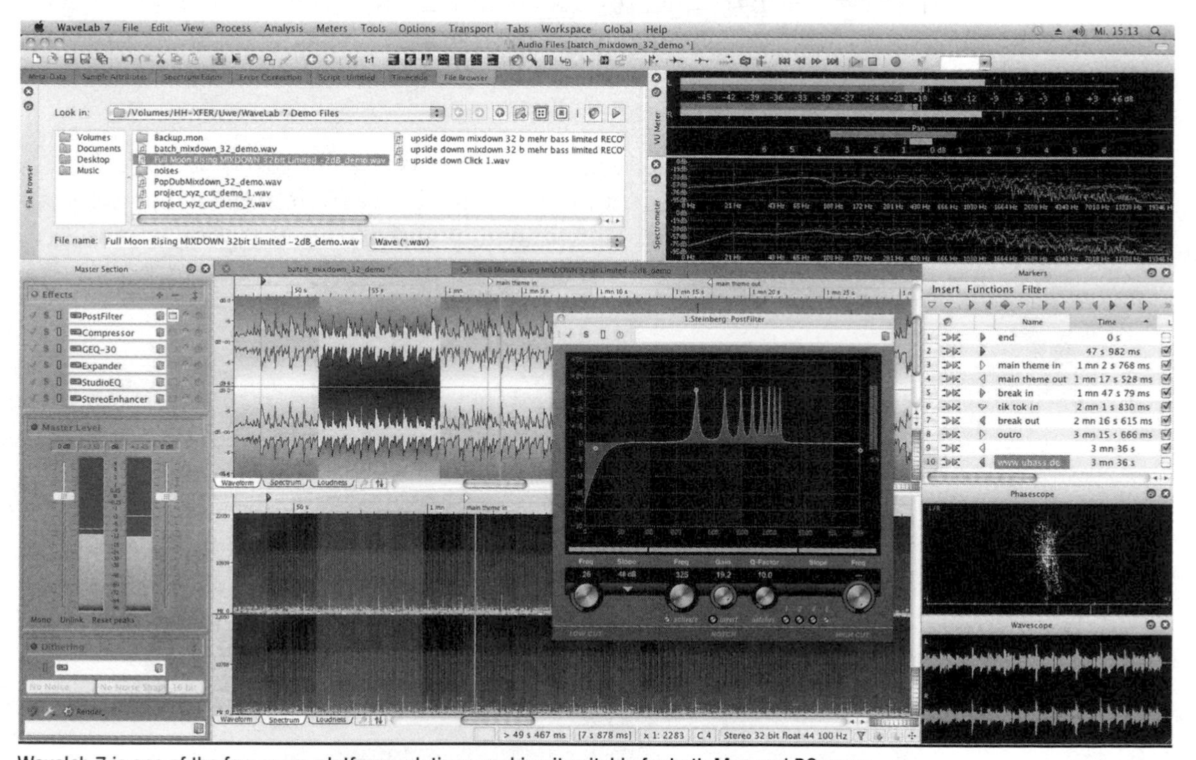

Wavelab 7 is one of the few cross-platform solutions, making it suitable for both Mac and PC users.

Steinberg wavelab

Like Sony's Sound Forge Pro 10, Wavelab 7 is a form of 'Swiss army knife' for audio, with a range of tools covering audio editing, audio restoration and sound design, as well as support for Steinberg's own VST plug-in format. As part of this suite of features, Wavelab also includes CD Mastering, with the option to export your final creation in DDP format. For some time Wavelab was only available on the PC, but Wavelab 7 finally sees the system evolve into a true cross-platform solution, making it the only option here that works with both Mac and PC.

2.8 Mastering plug-ins

While the provision of dedicated mastering software is somewhat thin on the ground, the situation for dedicated mastering-grade plug-ins is much more healthy. In truth of course, this is driven partly by the fact that many professional mastering engineers use a combination of both dedicated hardware and plug-ins for their signal processing (arguably using a 'best of both worlds' approach), but it's also great in that project studio users don't have to invest in £3,000 compressors before they can even think about mastering using their existing setup.

Although not exhaustive by any means, here's our pick of some of the most widely used and popular choices for mastering plug-ins. Beyond these, a range of fantastic tools exists, many of which we'll be highlighting in some of the more practical parts of this book.

iZotope Ozone

Mimicking popular multiband mastering hardware from the likes of TC Electronics, iZotope's Ozone is a great all-in-one plug-in solution for mastering signal processing. Ozone features a total of eight signal processing sections – maximizer, equalizer, multiband dynamics, multiband stereo imaging, post equalizer, multiband harmonic exciter, reverb, and dithering – each assessable from different software pages. It's a precise tool, with the level of flexibility you'd expect from a dedicated multiband signal processor.

Metering: peak v. RMS

Accurate and informative metering is an essential tool for any mastering engineer, giving them the means of assessing both the absolute electrical amplitude of a track and, to some extent, its perceived loudness. On the whole, volume metering falls into two broad camps – *peak* and *RMS* – both of which are important for a mastering engineer to track.

A peak meter reading measures the instantaneous electrical level of an input, ensuring that short transient sounds don't exceed the 0dBFS limit in a digital system. Although peak readings are essential for avoiding digital distortion, they give little or no information with respect to how our ears perceive music, which tends to work more with averaged signal levels rather than instantaneous peaks. An averaged meter reading – usually referred to as RMS, or root mean square – gives a much better indication as to its perceived loudness, allowing the mastering engineer to apply tools such as compression and limiting (which reduce the overall dynamic range) in a more informed way.

Given a need to assess both peak and RMS levels, most digital meters use a combined approach that shows both figures, as exemplified by a Dorrough meter. On a combined meter, the peak figure will always be higher than the RMS value, often using some form of peak hold so that you can see the maximum level that's been reached. The reading of both levels – peak and RMS – as well as the discrepancy between them, provides you some understanding of both the volume and loudness of your master.

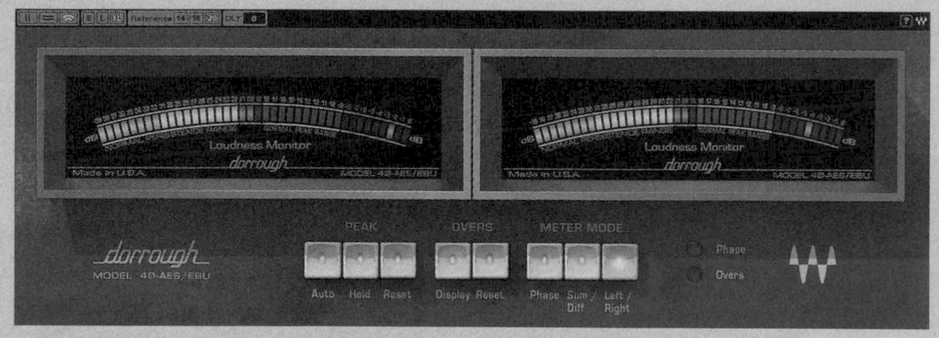

A balanced approach to meter reading involves an analysis of both Peak and RMS signals.

The current version is available into two flavours – Ozone 5 and Ozone 5 Advanced. The advanced version principally differs in that it provides an enhanced set of metering options, as well as individual plug-ins for each of Ozone's eight signal processing sections (rather than being combined in a signal plug-in).

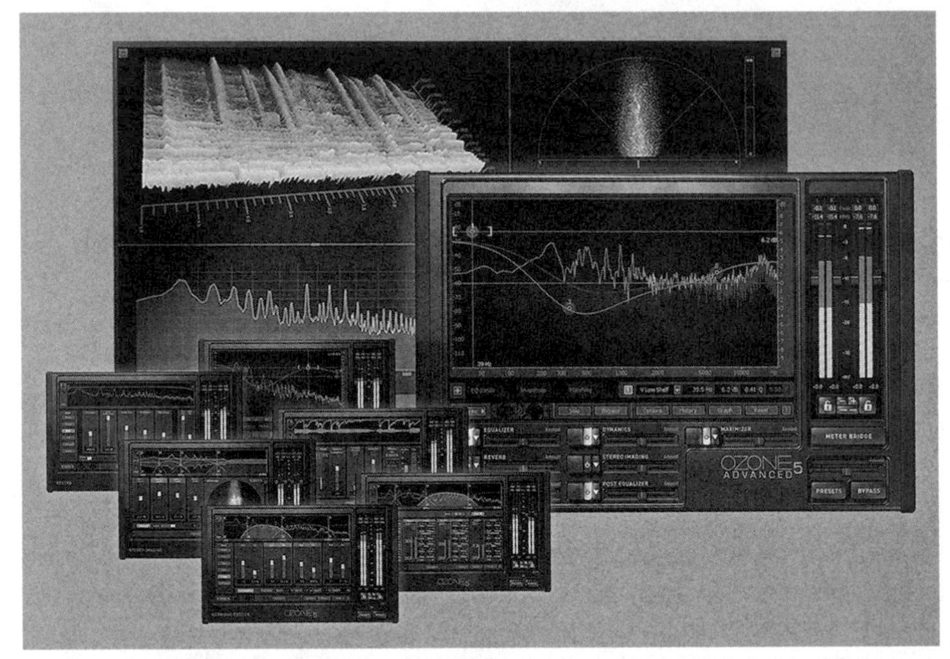

Ozone is a great all-in-one plug-in that contains all the key components of a mastering signal path – from equalisation, through to multiband compression.

Universal audio precision series

Universal Audio's UAD system has proved itself to be one of the most exciting platforms for plug-in development in the last few years. Technically speaking, the system uses a DSP farm card to

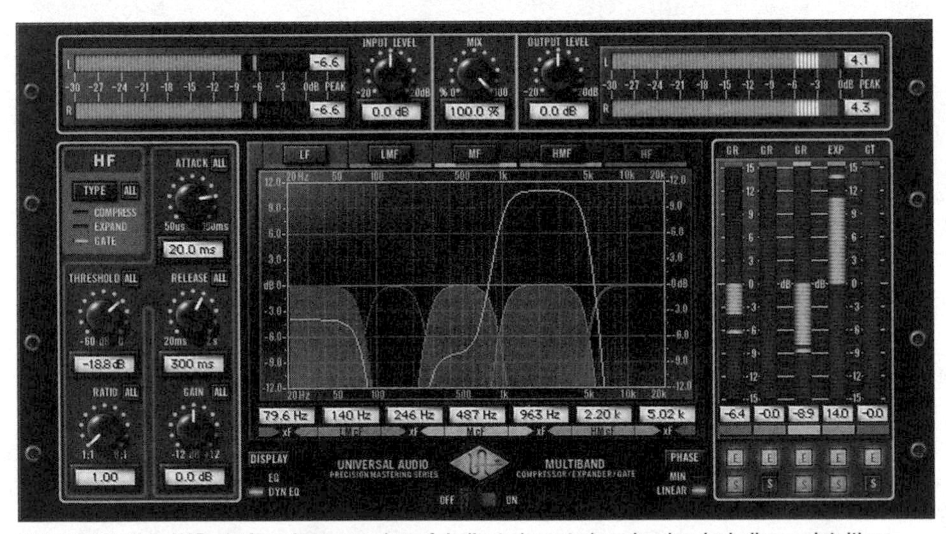

Universal Audio's UAD platform has a number of dedicated mastering plug-ins, including an intuitive multiband compressor and Bob Katz's K-Stereo Ambience Recovery.

power the plug-ins – either working over a FireWire connection, or by using a PCIe card installed inside a desktop computer. Overall, there's a range of plug-ins to choose from, many of which have been modelled directly from classic analogue hardware, including Studer reel-to-reel recorders, Fairchild compressors and Manley equalizers. The Precision Series has been specifically designed with mastering applications in mind and features key tools such as equalization, multiband compression and limiting.

IK multimedia T-RackS 3

As with Ozone, T-RackS 3 was designed as an all-in-one plug-in solution for mastering signal processing. There's both a standalone version, which negates the need for DAW and allows you to load audio files directly into the application for processing, and a plug-in version, which allows you to instantiate the T-RackS 3 processing as

Like Ozone, T-Racks 3 is a good all-in-one solution, although this time with a distinct bias towards modelled recreation of analogue hardware.

part of your DAW's workflow. Overall, there's a distinct 'analogue-like' methodology to T-RackS 3, with many of the plug-ins mimicking classic analogue signal processors such as the Fairchild 670 compressor and the Pultec equalizer.

Waves L-316

There are plenty of plug-ins we could pick from Waves's line of products, although it's arguably the limiters that are the pick of the crop in relation to professional mastering, with the L3 limiter becoming a ubiquitous part of mastering signal paths. When it comes to digital brick-wall limiting, the L-316 is one of the most powerful tools around, especially given its ability to apply limiting across 16 discrete frequency bands, unlike conventional broadband limiters. In theory, the L-316 negates some of the unnecessary modulation effects that can be produced by a broadband limiter, as well as allowing you to achieve a precious few extra decibels of loudness!

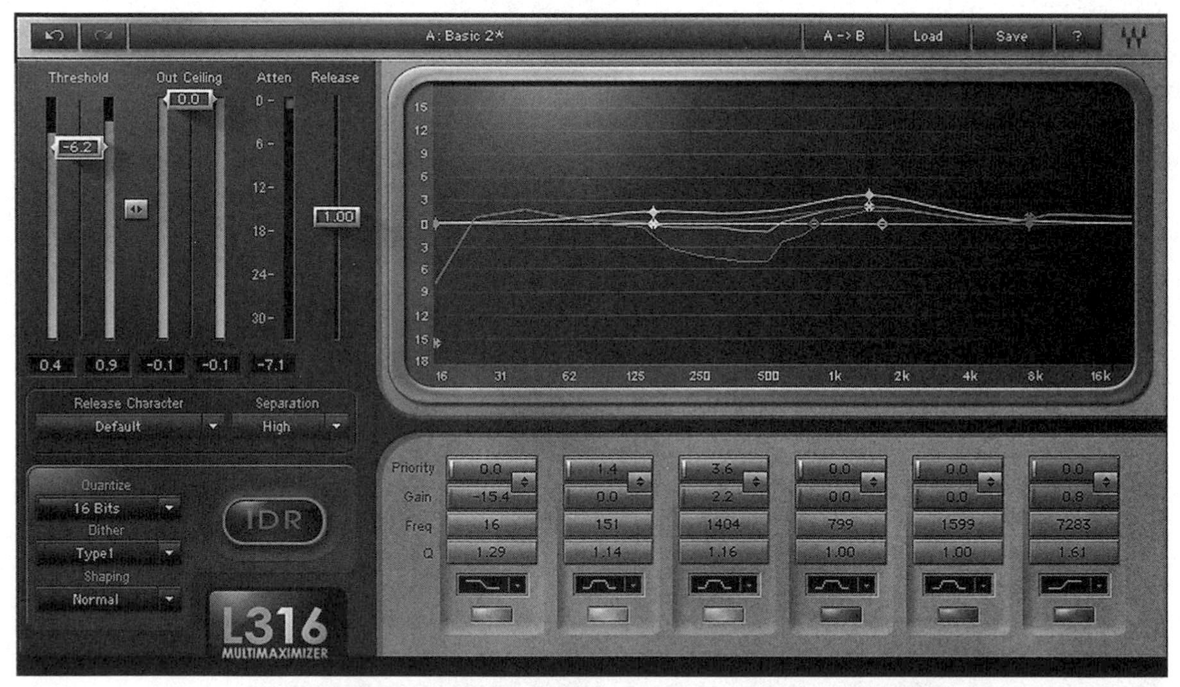

Waves' excellent plug-ins have gained an enviable reputation, but it's their limiters that get most usage in mastering studio around the world.

2.9 An introduction to M/S processing

Mid/side (M/S) processing will feature as a continual theme throughout this book, so it makes sense to introduce some of the concepts in this chapter. Although M/S processing is less important in music production circles, M/S tools form an essential part of mastering practice, and originates from the practices of cutting music to vinyl. Nowadays, M/S touches on all aspects of signal processing in mastering – from equalization to compression – as well as being an essential tool in how we refine the stereo image.

To understand the concept of M/S, we first need to look at the three principle ways a mastering engineer can dissect a given track. The first and most immediate dissection tools are, of course, the left- and right-hand channels of a conventional stereo signal. Mastering equalization, for example, might differentiate between the two channels of an L/R stereo mix – perhaps applying slightly more highs to the left-side than the right. Another dissection tool is multiband processing, where a two-track stereo mix is divided up into three or four different frequency bands. With multiband processing, an engineer could decide to apply more compression to one frequency band than another – shaping both the timbre and dynamic qualities of the end master.

Dissecting a track through M/S processing is essentially the 'third dimension' of signal processing, but to understand M/S we need to understand a new way of looking at stereo. As we know, stereo playback is a neat audio trick, using two speakers to create an 180° soundstage. What's particularly interesting about this two-speaker setup is that the centre of the soundstage is created by the two speakers having a degree of 'shared' information. In short, a signal fed at equal amplitude to the left- and right-hand speakers creates a 'phantom centre' image, almost as if there's a third speaker positioned equidistant between the two speakers. In effect, the *mid* signal is the *sum* of the left and right channels.

If the 'middle' of the soundstage can be expressed as the sum of the left and right channels, it logically follows that the two extreme 'sides' are formed by the difference between the left and right channels. Now we're looking at the stereo image in new way – formed from the *sum* and *difference* between the two channels – with discrete controls for the mid and side of the master. By controlling the master using mid and side controls (rather than left and right) we yield new levels of flexibility, so that we can differentiate between the EQ and compression applied to instruments in the centre of the mix (such as vocals) and signals that exist at the sides (such as cymbals overheads, reverbs, and so on).

2.10 M/S tools

Given an interest in the M/S dimension, there are a number of ways that you can integrate M/S processing into your setup.

Using stereo plug-ins in M/S format

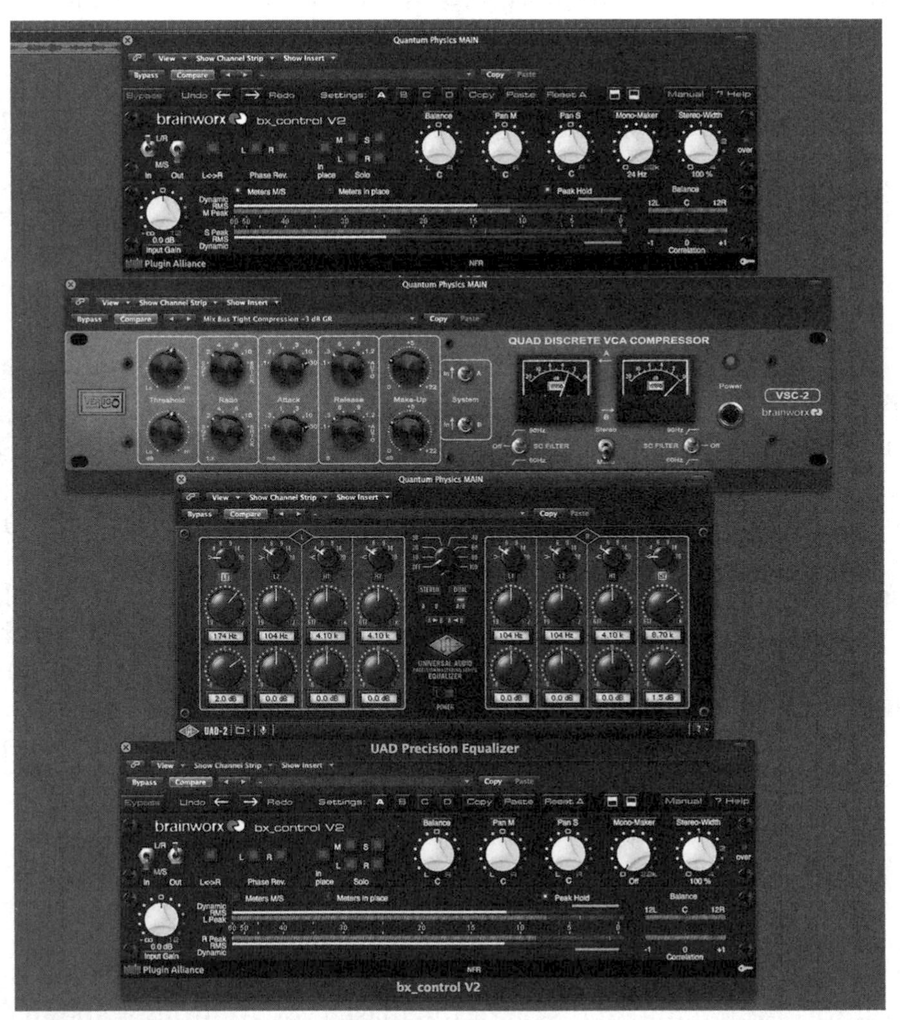

Conventional L/R plug-ins can be used for M/S processing by placing an M/S Encoder/Decoder at either end of your signal path.

To move between L/R stereo and M/S stereo all you need is a simple encoding/decoding plug-in, which performs the necessary phase tricks to convert the left and right channels into mid and side components. Place the M/S encoder – for example, Brainworx's bx_control V2 – as the first device in your signal path. The encoder converts the L/R signal into mid and side channels, with subsequent plug-ins in the signal path now processing the M/S components rather than a traditional L/R. Of course, if you're using a software compressor or equalizer you'll need to ensure that it has two discrete channels (rather than linked operation), so that you can process each channel in a different way.

Once you've finished your processing, you'll then need to use a decoder plug-in at the end of the signal path to convert the M/S signal back into a L/R signal. As part of that process, you might also get some additional controls (depending on the decoder plug-in) to control aspects such as stereo width.

Dedicated M/S plug-ins

A number of plug-in developers now create dedicated M/S plug-ins, or plug-ins with M/S modes that allow you to process a signal in the M/S domain without having to use the encoder/decoder combination. Good examples include Ozone 5, Waves' Center, Universal Audio's Fairchild 670 (which models to original lateral/vertical M/S systems used for vinyl mastering) and various offerings from Brainworx. Dedicated M/S solutions often offer a number of additional easy-to-use features that are a product of the M/S conversion, most notably Brainworx's Mono-Maker feature, which creates a tight mono image to the mix below a given frequency.

In the following chapters M/S will reappear several times, and we'll start to see how these unique tools can be used as part of the mastering process.

Visualizing your audio

As well as measuring the amplitude of a signal, there are a variety of other metering tools that a mastering engineer will turn to as a means of better understanding the auditory qualities of the finished master.

Spectral, frequency-based analysis uses *fast fourier transforms* to break a signal down into its individual harmonics, showing the relative amplitude as you move throughout the entirety of the audio spectrum. Often attached to an equalizer plug-in, spectral analysis offers both an understanding of the overarching timbre of your master, as well as specific frequency issues within it – whether it's an unwanted bass resonance or some mains hum.

A phase meter shows the correlation between the left- and right-hand side of a stereo signal, providing some indication of how the music might translate if both sides of the stereo image were summed to create a mono signal. Where the correlation is good – with a reading on the positive side of the scale – mono compatibility is assured. With a negative reading, some degree of phase cancellation would occur if the signal were collapsed to mono.

A stereo vectorscope provides some indication as to a master's stereo properties. A mono signal produces a single vertical line, whereas divergence between the left and right channels creates a form of dancing Lissajous figure. With a little experience, you can extract a lot of information from a vectorscope, even down to the individual panning of single instruments and the use of stereo effects such as reverb.

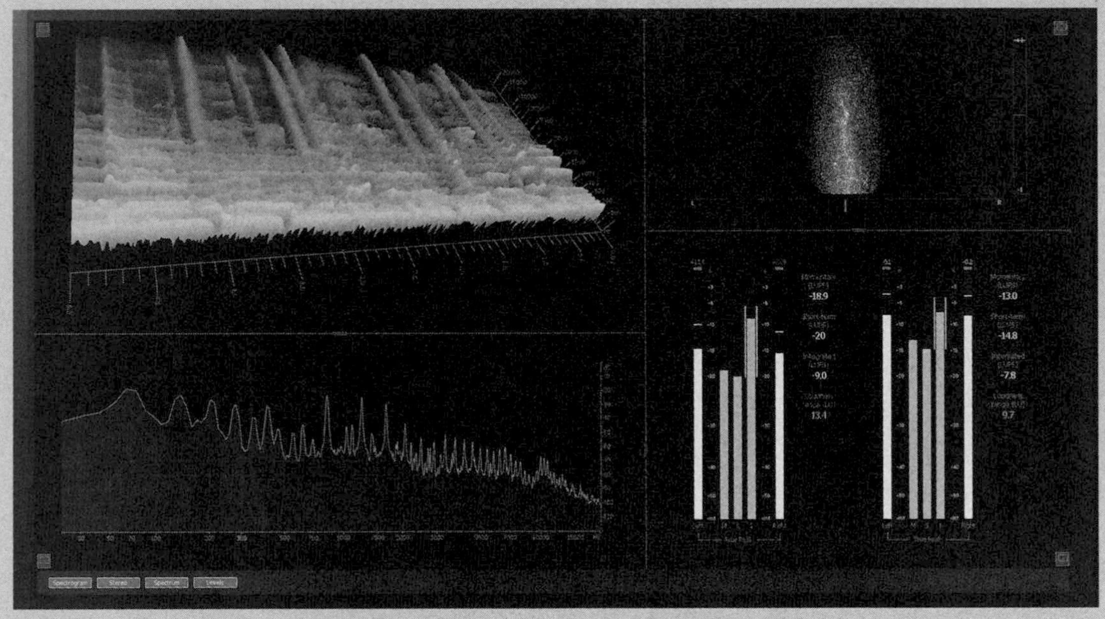

As well as sound pressure levels, there are a variety of ways you can meter your finished master.

Dedicated M/S plug-ins offer an easy way of experiencing the wonders of M/S processing.

2.11 Essential improvements for project studio mastering

While it might not be practical or financially viable to completely re-equip your studio for mastering work, there are some areas worth investigating, where a little extra investment and thought can make a big difference to your sound.

Acoustic treatment

Most project studios suffer from poor acoustic treatment, which ultimately impinges on the decisions that you make when you're mastering. Reflections bouncing of nearby walls can colour the sound at your listening sweet spot, as well as clouding the soundstage and affecting your ability to localize sounds in the mix. Standing waves can also create unwanted peaks and troughs in the bass response, so that what might appear initially bass-light is actually bass-heavy, and vice versa.

Affordable options for acoustic treatment include a number of user-installable kits that feature a range of absorption and diffusion panels that help control reflections in your control room. On the whole, most of the acoustic treatment is focussed on the 'transmission end' of the control room, with absorption panelling behind the main monitors, as well as either side of the listening sweet spot to capture and divert the initial early reflections. Treatment at the back of the room is slightly less severe, often using diffusers to scatter the echoes, rather than absorbing them all. Finally, bass traps in the corner of the room help control bass resonances, making the bass end more unified and distinct.

Monitoring

Near-field monitors aren't ideal for mastering, so if you do have to use pair of small speakers, ensure they're produced by one of the more reputable manufactures, such as PMC or ATC. As we saw earlier on, near-fields are principally inhibited by the fact that they use two-way drivers, missing the vital midrange 'squawker' that delivers some important information about the timbral balance of our music. One interesting new direction, though, are a number of three-way monitors – including ATC's SCM25 – that are specifically designed to work in smaller control rooms, and arguably provide a more balanced tonal picture of the music you're trying to master.

Adding a sub is also another beneficial investment, as this will improve your monitoring in respect to the low-end, which tends to roll-off sharply on a pair of near-fields. Decoupling your monitors – using something like Primacoustic's Recoil Stabilizer – can also improve transient response, imaging and the tightness of bass.

Unless you can afford big three-way monitors and expensive custom-designed acoustic treatment, it's best that you run your monitors in a near-field position – in other words, arranged just a metre or so in front of you forming an equilateral triangle. Positioned in this way, you get a good presentation of the stereo soundstage and avoid too much room colouration and flutter echoes from clouding your judgment. Also, get to know the 'voicing' of your monitors – seeing how different styles of music translate across your 'perfect' listening environment.

Conversion

Although you might not be venturing out into the analogue domain, the use of conversion is still a vital part of mastering in a project studio environment. Even if you process your master using 'in the box' plug-ins alone, the quality of the D/A conversion is vital to give you an accurate picture of what you're achieving. Principally speaking, poor-quality conversion tends to deliver an unwanted colour to the sound (which might inhibit your application of equalization and other timbral tools), as well as a lack of detail and focus in the soundstage caused by 'jitters' in the sample clock. While rectifying a small issue such as clock jitter might not seem worth the effort, it will affect your ability to assess finer details such as compression release times, as well as the depth and definition of your soundstage.

On a positive note, the quality of conversion on most project studio equipment has seen some major improvements over the last few years, especially in respect to companies such as Apogee producing ever more affordable products. In truth, therefore, it might be that one of the more 'audiophile' FireWire and USB audio interfaces delivers as much clarity and detail as any dedicated mastering-grade converter. Ultimately though, investment in good quality converters will pay dividends in the long run, as well as making your working day proportionately more enjoyable!

CHAPTER 3

Mastering Objectives

In this chapter

3.1 Introductory concepts 38
3.2 Flow 39
3.3 Shaping 40
3.4 Listening 41
Listening interface and quality benchmark 41
Knowledgebase 42
Listening subjectivity versus objectivity 43
Holistic, micro and macro foci in listening 44
Listening switches 45
3.5 Confidence 47
3.6 Making an assessment and planning action 48
3.7 Assessing frequency 48
3.8 Assessing dynamics 49
3.9 Assessing stereo width 49
3.10 Assessing perceived quality 50
3.11 The mastering process 51
3.12 Capture 53
3.13 Processing 53
3.14 Sequencing 54
3.15 Delivery 55

Having explored the principle tools that power the mastering process, we now need to consider the important fundamentals that form the backbone of what could be considered 'effective practice'. While it would be tempting to jump straight into the practical process of mastering – adding compressors and equalizers, for example, to make our music appear subjectively better – it's worth taking some time to consider the important conceptual building blocks of the process we're engaging with. Like song writing, recording or mixing, mastering has its own set of objectives, as well as a unique methodology and workflow that marks it out as being distinct and separate from the rest of the production process.

Some of the important factors that we need to consider in this chapter include the overarching criteria of an effective master: addressing concepts such as flow, for example, as well as the needs of shaping and developing a listening experience over 45 minutes or so. Even more importantly, we'll take a detailed look at the process of listening – something that's easy to take for granted, but is actually a skill that empowers every action a sound engineer engages with. Whether it's the ability to switch between a macro/micro focus, for example, or being able to critically assess the properties of the music you're trying to master, listening will always be the key to effective, professional-sounding results.

Towards the end of the chapter we'll also define what we consider as being the key stages of a typical mastering workflow – something that we touched on in Chapter 1, but at this point, something that needs to be developed so that we understand the entirety of the mastering process. All of this forms the backbone for the subsequent chapters that step through sonic concerns such as timbral balance, dynamics, stereo width, loudness and, of course, the process of editing and assembling the final master ready for replication.

3.1 Introductory concepts

We'd naturally suggest that mastering is essential for whatever musical project you're completing. It provides objectivity and a final creative stage before release to the wider world. In the current climate of Internet-driven distribution, it is understandable that single tracks are released or uploaded by bands, labels and less legally-abiding individuals. Mastering is just as important on a track-by-track basis as it is for the whole of an album. Despite discussing much of the following in the context of an album, the information, on the whole, will be of relevance to single-track masters too.

Understanding the role of mastering makes the album that you buy in the shops rather more than that collection of songs someone has written and recorded. Of course the mastering engineer could just collate the 12 tracks together and release them as an album for you to buy. However, I dare say that some of the most compelling albums of the last 50 years may not have seemed so 'complete' had it not been for the mastering process. Therefore is it important to consider at all stages the importance of the Master and what its end purpose is for.

A great deal of time and effort goes into the composition, recording, production, mixing and ultimately the mastering of an album. Within mastering there are strategic levels and ways of considering how to approach a product. The producer, musician and anyone else with vested interests in the final product will have their own take on this process and desire for its outcome. These must be taken on board by the mastering engineer and be interpreted and reflected in the final copy. Yes, mastering can enhance and can ultimately bring the best of an album out to the customer, but the music, recording and the mix has to be there in advance otherwise there will be an uphill struggle to create the desired quality.

3.2 Flow

Consider your favourite album of all time, and take some time to look at the liner notes. Observe who mastered it and where. The mastering engineer in question has brought that collection of songs to you as a complete-sounding album. That was achieved by sculpting aspects of each mix individually, then taking those aspects and rounding them to blend a perfect album.

Flow is just what we refer to as a perfect outcome – an album that flows from start to finish. There are classic albums we all know and love which have this so-called flow. Imagine if one of the tracks was re-ordered. Would it flow in the same way? Perhaps not. This is all fairly subjective, as you've listened to and learnt the album, and so expect the songs to sound the way they sound and to start when they do. However, someone has made the sonic decisions that made this a reality. With the artist and producer's cooperation, this is the mastering engineer's forte.

Admittedly, flow can be lost now that many online digital download sites such as iTunes permit per-track purchases. In response, some artists even insist that their albums are sold as one purchase. Unfortunately, many misconceive this as being because they wish to gain higher sales and profit, and while in some cases this might be true, for others it is because they wish to keep the concept they've worked hard to create – the experience, if you like.

Achieving flow is all about an affinity for music and sound. The average mastering engineer can work on many albums per week, and in doing so will encounter a huge variety of musical styles. Having an appreciation of many styles and the ears to listen to what works and what doesn't will leave each mastering engineer with a personal blueprint of how to try to achieve flow.

The parameters for achieving this flow are set out in the coming chapters and discuss aspects such as dynamics, tonality and time. All of which make the ingredients that allow us to tamper with a collection of works and make them into an album. How you approach making what we call flow will be down to your experiences as a music listener and abilities as a mastering engineer.

We will visit the concept of flow again in Chapter 9 as we knit together all that we've learnt. Meanwhile, let's consider what it is a mastering engineer is trying to achieve.

3.3 Shaping

In their book *Last Night a DJ Saved My Life*, Bill Brewster and Frank Broughton (1999, Headline Book Publishing) liken a DJ set, with its highs and its lows, to that of a three-minute pop song. They relate the same discipline of shaping the emotional architecture of the three-minute pop song, with its crescendos as choruses within the piece and a breakdown with the middle-eight. The structure can be very formulaic.

Despite being only three minutes, what is known as emotional architecture can be shaped to provide the listener with a series of emotions or expectations. Brewster and Broughton conceives that a three-hour DJ set can achieve the same sort of shaping as that of a three-minute pop song but expanded to suit the DJ's set. Similarly, the producer, artist and mastering engineer have the ability to shape the emotional architecture of an album over the whole 40 or so minutes. How this can be shaped is something we'll explore more in this chapter and also in Chapter 9 as we look at the album format in context.

3.4 Listening

How we hear and interpret sound is extremely subjective, yet we need to quickly discuss this as it's the essence of the decisions we make, whether we're recording, mixing or mastering. In this section we briefly explore some aspects of listening and some personal concepts of how you might analyze what you're listening to.

Listening interface and quality benchmark

In any audio work, whether that be recording, mixing or mastering, it is important to understand what you hear. The monitors you use can of course colour the sound and give you a slanted view of the audio you're hearing. The same slanted view can be affected by the room in which you're listening and numerous other areas such as your convertors.

What all audio professionals, and mastering engineers especially, should be aspiring to is a near-perfect monitoring environment in which objective judgement is intact. However, to achieve perfection can be be terribly expensive, although it is something to aspire to. In the interim, there are numerous things you can do to improve both the acoustics of your room and the monitoring quality (which we'll not cover here).

When optimized, your monitors and environment should provide you with a trusted invisible connection to the music – your listening interface. You're developing a confidence that what you're hearing is actually what is happening and not a byproduct of your room or monitors. Once you're intimate with the sonic quality of what you're listening to, it is important to consider something we refer to as our *internal quality benchmark* – a threshold of quality by which you measure the sound quality of the music you're listening to.

How you achieve this is to listen to and assess as much high-quality audio material as you can through your monitors, in the optimum listening position. How loud you listen engenders considerable debate, with some (such as Bob Katz) suggesting that your monitors could ideally be at a fixed level as with cinema sound, to those who mix and master at very low levels. Decide on the level you wish to monitor at, and perhaps try to keep it there, resisting the temptation to turn it up as you enjoy something you hear. This can help to achieve a level of objectivity, or disconnection from the music, whilst turning your attention to the audio quality.

Knowledgebase

File formats, dither (word length reduction) and sample rate conversion

As you begin mastering you will need to decide on the file format and specification you wish to work in. This, to some extent, will be dictated by the DAW you're using. Some, if Mac-based, will perhaps prefer to use the Audio Interchange File Format (.aiff), although this is less prescribed these days. Meanwhile, those using PCs might be more at home with .wav files.

To some extent, on many DAWs the file formats can be interchanged or converted with no loss in quality. However, you will have quite a lot of choice in specifying the sample rate and bit depth. Before spending a great deal of time discussing each one in turn, let's consider the outcome first.

Early on in digital mastering the purpose was quite simple – to master to 16 bit and 44.1 kHz, ensuring that the material meets the needs of the distribution format of the time – CD. As such 16 bit, 44.1 kHz audio has become a base standard for what we call PCM audio. PCM actually stands for *pulse code modulation* and describes the way in which the data is written onto a file format. However, PCM is widely used as a term to express uncompressed audio data as opposed to the data compression formats such as MP3 and AAC.

So, looking at a lowest common denominator view, it would be best to consider a session on your DAW running at 16 bit and 44.1 kHz as standard. This will mean that the album will work seamlessly when bounced for a DDP, with the eventual aim of making it into a CD.

Currently however, two problems upset this simple balance. Firstly, for many good reasons recordings can, and should, exceed the specification of CD, usually with higher bit rates and in many cases higher sample frequencies. Secondly, there are often issues relating to your deliverables. Sometimes clients will want lower-quality audio (data compressed MP3 and AAC) and may also wish to retain the 96 kHz and 24 bit audio for a higher resolution release later down the line. This is loosely the current concept behind Mastered for iTunes.

Considering that 16 bit and 44.1 kHz is still just about the most likely specification, there may be times when it is necessary to 'reduce' the quality of the audio to meet with the final file format. In doing so two processes will be required which ought to be mentioned in a book on practical mastering: dithering down and resampling.

Listening subjectivity versus objectivity

Apart from the specialist equipment and enhanced monitoring environment the mastering studio provides, the reason the mastering engineer is so important to the music production is the objectivity he or she brings to the project. Many a positive relationship has been made between producer/engineer and the mastering engineer. Tracks can exchange hands long before they come to the mastering engineer for mastering itself. Engineers will send over material if they want an unbiased opinion on what they're working on, and the mastering engineer will be happy to give it.

This requires a high level of objectivity and is the reason why mastering engineers (and ever more so today, mix engineers) are separated from the lengthy and involved writing and recording process. This can be the first stumbling block for those wishing to master their own material, as there can never be full objectivity once the first finger has been put on the piano key to begin the writing process. This is not to say this does not happen, but in our experience it is rare.

Developing objectivity is not something that can be achieved easily, and it involves the consideration of how you actively listen to the music you work on. It is very easy for one to listen to a piece of music from a client that you really like and for this fact start to cloud your assessment of it. It may seem clinical to assess music this way, but it is to permit you to listen in an objective manner and make judgements, such as how much bass should be on this record, given that many people will be listening on earbud headphones, or inexpensive hi-fi systems (with added bass effects!). Whether you like the tune or not doesn't come into it (although it certainly can be fun).

Holistic, micro and macro foci in listening

Mastering can be thought of on many different levels. At an intense level it can be considered as changing slight nuances, adjusting tonality and microscopic elements of a piece of music all the way through to the wider concepts that make an album which might be segued over 45 minutes.

One concept to consider is what we call the *holistic, micro and macro foci* of the music, as discussed in *What is Music Production* by Hepworth-Sawyer and Golding (Focal Press 2011). This builds upon work from Katz and others and discusses the way in which we can actively or passively listen to music.

In essence, the concept assumes that listeners are either passive or active. An active listener would be concentrating on the music at hand, whereas someone working or cooking with the radio on in the background would be considered a passive listener – passive not just in terms of their attention to the music, but the listening position and acoustic environment.

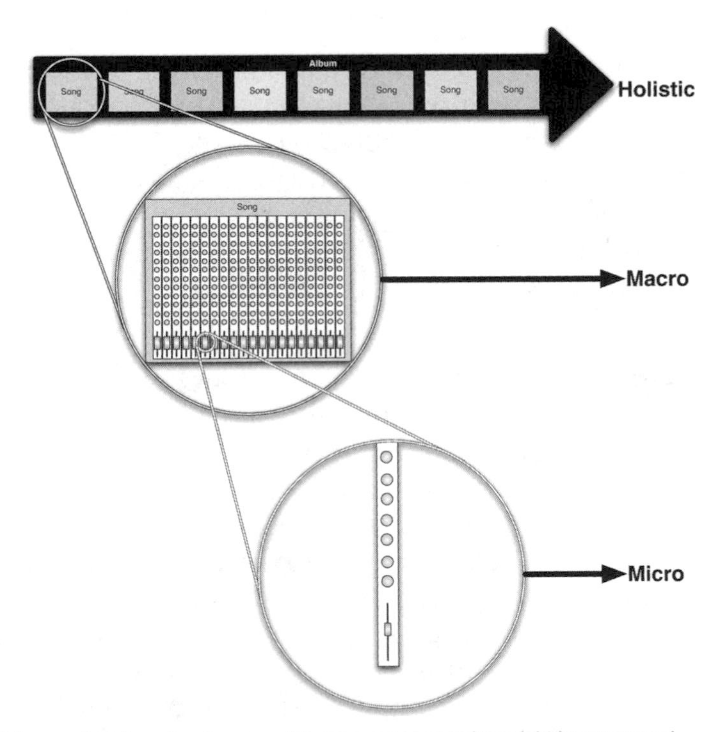

Listening styles can be interesting to appreciate how your mastering might be consumed.

Across this is also the added dimension of whether the listener is considering the whole of the sound, in this case a macro focus, or is listening intently on one aspect and concentrating on the micro. It could be argued that a keen guitarist emulating a guitar hero is more likely to focus on the guitar than on the whole, and could be therefore considered a micro listener. In contrast, a mix engineer can zoom out to work in the macro focus too, taking in the whole of the music as necessary and then switching into the micro (emphasized by the use of a solo button perhaps) to make repairs or edits to certain aspects of the mix.

This micro and macro listening model is intended to be thought of within the confines of a single song. The mastering engineer has to be able to switch between these two modes on a song-by-song basis, but will spend more time in the macro, listening to almost everything all at the same time.

This concept also embraces what is known as 'holistic' listening. This is intended to reflect the mastering engineer's traditional role of zooming out further and considering the whole of an album and making edits accordingly, within the micro and macro foci.

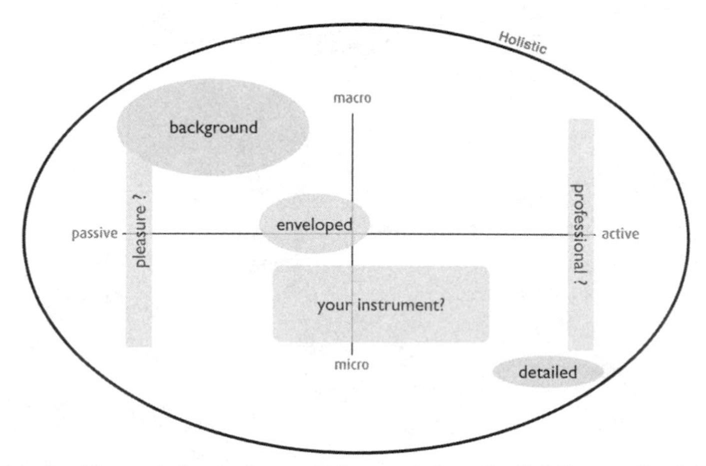

The foci of Listening: The mastering engineer might concentrate on the Holistic as well as both Micro and Macro.

Listening switches

Another concept is the ability to switch modes of listening from micro, to macro, to holistic depending on the work you're engaged in or the requirements of your task. Additionally, it is worth considering the way in which the material you're working on will be listened to both passively or actively.

A so-called *listening switch* can be a useful concept to keep handy. Imagine if you could engage and disengage your interest in the music you're listening to – a subjective/objective switch, if you like. It is worth considering that much of the listening public probably focus in on singing and the melody, paying much less attention to the backing. However, a music professional will either concentrate on the whole together or certain aspects across the arrangement.

Consider how you listen to a new single you've bought and how you become familiar with the music. Is it the vocal hook that has gained your attention, or is it a culmination of a number of factors? Try listening to the music both objectively and subjectively, and then trying to remain in a macro state listening to everything at the same time, perceiving levels all at the same time in different frequency bands, then in micro focussing on a particular element (without the ability to solo it!).

Dithering

Imagine a simple analogy whereby you're being asked to post balls through the open sections of a ladder. The ladder's gaps are essentially the quantization points of the digital audio system. In other words, when a ball gets through the gaps then the converter can measure the amplitude of the signal and move on. What if the balls hit the rungs, bounce back and don't go through to the convertor? Well, they'd not get measured, or quantized, and the digital audio stream would contain gaps.

Dither is actually a quiet noise that is added to a signal to assist in the quantizing process of the ADC. Within this analogy the noise could be considered as something that makes the ball a little agitated, say a bit of a gust of wind. This agitation would be enough to push the ball above or below the steps of the latter and through the gaps – goal!

The above analogy may be far fetched, but it is a simple one to try to understand the process in layman's terms. In the real world however, dither is a little bit more involved. Dither is needed because lower bit rates are not as resolute as, say, a 24 bit file, and as a result it is quite possible that those waveforms that do not meet the magical quantization levels (the gaps in the ladder) could be lost completely and threaten the accuracy of the conversion process.

Dither therefore somewhat blurs this and permits every ball to reach the most appropriate gap in the ladder, thus applying a best-fit approach. This works for all those bits of audio that are recorded with an analogue-to-digital convertor, and will be of importance to you as you capture audio from other analogue sources, such as a turntable or open-reel tape machine.

3.5 Confidence

One aspect of mastering is your confidence in the tools you have, especially the monitoring environment you possess. Many recording engineers miss the prevalence of the Yamaha NS10M Studio Monitors, as they hold a near-iconic place both in mixing history and on the meter bridge of most recording studios. The NS10's as they were known, provided a voicing that no other monitor quite managed and became a standard among mix engineers in the 1980s and 1990s. They were, as loudspeakers go, not that great by today's standards, with limited bass response, and a voicing which favoured the upper-mids.

What made the NS10's so popular were perhaps two things. Firstly, the age-old adage of 'if it sounds good on NS10's it'll sound good on anything' simply works. They made you work hard at mixes, and were very quick to permit you to bed the vocals in just right, whereas larger monitors would somehow give you so much bass and treble that you'd feel the need to push the vocals too much to compensate. The second benefit was the ability to take these relatively small speakers to any studio that did not have a set (or more likely, every studio had a set anyway). Add this to the fact that you would be working in the near field and you'd negate many of the acoustic effects of the studio, leaving you with a very even monitoring environment wherever you worked.

This meant one thing – that engineers got used to a reference monitor and began to know it well, developing their *listening interface.* Despite the NS10 being completely unsuitable for mastering, the moral of the story is that once you're used to your monitoring environment, you'll develop a similar confidence around what you hear, which will allow you to make accurate judgements about the music you are working on.

Confidence doesn't just stop with what you hear, but what you'll do to what you hear. Knowing your tools, as we explored a little in Chapter 2 and will follow in more detail in the coming chapters, is equally important. Understanding the finer details of dynamics processing, timbral modification and other features available to you will improve not only your agility as a Jedi mastering engineer, but also ensure you make the right decisions more often.

After becoming one with your monitoring environment and equipment, it is also important to consider the way in which you'll assess and approach the mastering of each track you're faced with. How will you assess what needs to be done to any given track once you've listened to it? Given how you've assessed the audio, how will you then use the tools at your disposal to create a finished master?

Developing confidence as a mastering engineer will take time, as you experience many different projects and many different styles of music. You will approach each track differently and develop the appropriate skills as you go along, building a library of tools and techniques that you can call upon depending on the task in hand.

3.6 Making an assessment and planning action

With the groundwork completed you should have developed a quality benchmark and be situated within a solid, and familiar, listening environment. As you receive each piece of music you will need to make some initial assessments based on a number of audible factors, and this will permit you to plan a course of action, employing the skills you'll be introduced to in Chapters 4–8.

It is interesting to note that some people visualize colours as they listen to music. This provides guidance for them to assess sonic quality (imagine something similar to the graphics iTunes creates as you listen to music). Meanwhile, others we've noted tend to analyze the music against frequency as they're listening to it. Think of a spectrum analyzer, or even a Fourier transform graph, where each frequency band is instantly appreciated and analyzed against its level. This is a very helpful way of hearing sound for a mastering engineer, as one can easily identify issue areas in a mix and what might be a course of remedial action.

3.7 Assessing frequency

It is therefore necessary to assess something you're listening to for the first time against its frequency content. Sometimes you'll get a bright mix and sometimes you'll have bass-heavy content to deal with. You'll only be able to analyse this if your personal quality benchmark, against which you're making your assessment, is accurate.

If the material is too bright then you will need to make an assessment as to whether this is just a spike in the audio material or something more generic such as a lack of bass in the whole mix. The former can be detected from an overly hyped cymbal or a presence peak accentuated on a vocal track. However, the latter could be caused by a bass-heavy monitoring environment during the mix (quite common today with badly incorporated subwoofers). We'll explore this further in Chapter 5.

Similarly, you may wish to make improvements or timbral changes to an already good mix. Choosing the correct band to boost or cut will bring necessary enhancements to various parts of the mix, and will of course make changes to many of the instruments within the mix. Therefore it is of utmost importance that you take care in doing this, as you will not want to change the focus of the mix in any way. EQ will be one of the tools we employ to help correct or alter these issues. This is covered in more detail in Chapter 5.

3.8 Assessing dynamics

Similar assessments need to be made about the dynamics of the music you're presented with. Sometimes, regrettably, you'll be faced with a buss-compressed mix from a studio with little life left in it. Ironically, the compression applied is sometimes more than a mastering engineer might apply. That might be through better tools more adept to the job at hand, or better processes to achieve the same end. On the other hand, you might receive a really dynamic piece and need to employ some careful processing to give it a solid and deliverable form.

This could be classed as 'power'. The dynamics are, after all, the power of a piece of music combined with the overall mean loudness. Much discussion outside of this book is dedicated to this topic, so we'll not dwell on it right now, but be prepared to make appropriate assessments of dynamics or loudness according to genre and taste, and ensure that you appreciate those recordings which still contain dynamics and breathe.

3.9 Assessing stereo width

Many other assessments will be made as you come to learn about the sonic characteristics of music production. One such assessment will be the stereo width. Far too often we receive mixes that have quite a lot of phase issues, due to some aspects of the mix being incorrectly out of phase with other aspects (commonly sampler patches of pianos and so on being exceptionally wide, with limited mono-compatability). If we're speaking riddles here, don't worry, as we pick this up in Chapter 7.

Perhaps the mix lacks width, but has something you can exploit using some of the tricks up the mastering engineer's sleeve. Alternatively, if the mix is too mono there may be very little you can do, but you should check with the mix engineer that this was as the production team intended.

3.10 Assessing perceived quality

Far removed from the assessable factors listed above, you may also want to subjectively measure the music's sound quality using your quality benchmark. You will immediately appreciate if the work has been well produced or whether this is just good music with the faders thrown up.

Why should this matter? Well, if a well-produced album should come across your DAW, then you'll know how to handle it. Often, the production values will be imprinted into the message it provides because of a detailed and focussed mix. Hooks will be immediately obvious and will grab our subjective attention. These types of mastering jobs can be the easiest to respond to and master. You can simply improve the overall quality of the music and enhance the delivery of those important hooks. It is also worth remembering here that it is perfectly possible that you've done your job as the mastering engineer if you do absolutely nothing at all – if that's what the track calls for.

Alternatively, in our crude example above, should the faders have been 'thrown' up and not so much sculpting has taken place, it can be difficult to find the focus of the track and master accordingly. Yes, you can respond by ensuring nothing untoward is going to poke out of the listeners' headphones, but it must be noted that you cannot really fix a mix in the master, any more than you can fix a poor recording in the mix. Agreed, there are elements you can improve upon, but nothing will beat an excellent recording and mix being delivered to you in the mastering studio.

Questions should then expand to ask how that individual track fits in holistically with the album around it. This will have an impact on how the mastering will work in context. What might appear to be a single song which sounds unfocussed on its own might come sharply into focus when wrapped within a story and flow of an album.

Perceiving quality in music is naturally a subjective thing, but it is important to keep as much objectivity as you can at this point, assessing each individual track on its merits and also in context of a wider album (presuming it is being mastered in this context).

3.11 The mastering process

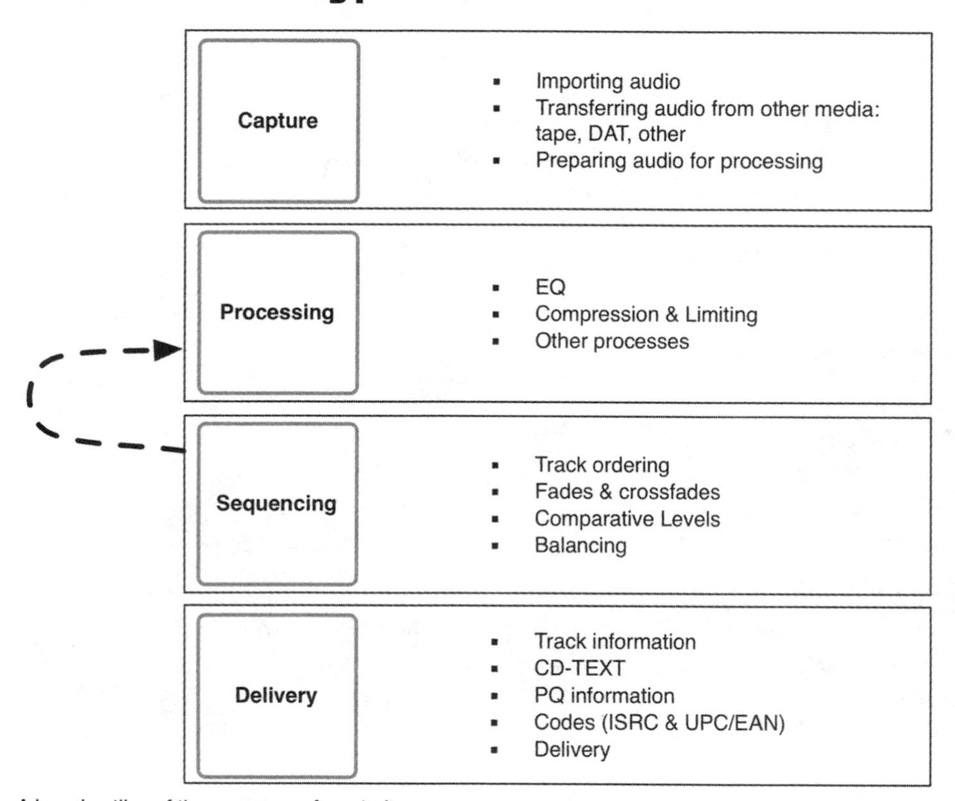

A broad outline of the processes of mastering.

In music production, an adopted production process leading from the pre-production stages all the way through to the mix and then mastering has evolved and has been accepted. Separating out pre-production from the studio recording sessions may not always be

Word Length Reduction (WLR)

If you're working with 24 bit digital files in the DAW, it is fine to continue to work at this higher resolution as this is the best quality you have, and as such you may wish to keep these files for higher resolution releases of the material. Despite this, for CD you'll need to get this extended bit depth down to the 16 bit specification. This process is known as word length reduction (WLR).

So how can we do go about achieving this? Well, the first and most obvious way would be to truncate the wordlength – essentially chop off the last 8 bits of the digital word. However, this would cause some significant losses in the quiet and important details that make up the richness of our audio. So intelligent methods are identified to lower the wordlength by adding digital dither methods to the signal. How these are achieved are outside of the remit of this book, and many resources in the Focal Press catalog cover this. However, what we must take from dithering down is its practical application.

Practically, dithering down audio material is a digital process which can, on the most part, happen automatically if you allow your DAW to take charge of the process. This is not such a bad thing, as our understanding and management of digital audio is much better nowadays. Nevertheless, it is worth being aware of dither, and the choices it presents to you may affect the quality of your audio.

Dither comes in various shapes and sizes (literally) – with a whole host of algorithms to make your resolute 24 bit file squeeze into 16 bits. The kind of dithering above describes its most basic forms, while noise-shaped dithers shift that noise to frequencies above 20 kHz (which we generally cannot hear), thus improving the performance.

The POW-r suite of algorithms is perhaps the most famous 'brand' of dither. It is employed by Waveburner and many other professional mastering products. POW-r cleverly employs psychoacoustic modelling for shaping its noise, and as such is said to have the most forgiving algorithms in the cupboard. For more information on dither and the noise shaping plotted on frequency graphs, please visit: www.weiss.ch/pow-r/documents/pow-r_brochure.pdf

iZotope's Ozone employs their own dither algorithm called MBit+ which has received some critical acclaim similar to that which POW-r received upon its release. The good thing here is that MBit+ has more features that allow the mastering engineer to adapt the dithering ever so slightly. Controls such as noise shaping pre-sets and dither amount (none, low and high) are included.

the best thing, but does happen frequently. Additionally, mixing appears to be something that is increasingly removed from the team that records the work. Modern mastering has always, and should be, a separate stage in the process due to the objectivity it can provide for a project.

Within mastering itself there are standard processes that engineers go through to bring a piece of music to market. We outline these broad stages here and in doing so establish the basis for the following chapters.

3.12 Capture

Whether it be a data file transferred via FTP or one of the newer large-file services such as yousendit.com, or an older format such as Digital Audio Tape (DAT), the audio needs to be imported or captured into your DAW – the hub of your operations.

In Chapter 2 we discussed the need for a good analogue to digital converter, and the ability to capture this correctly from the playing machines. For example, if you're capturing analogue sources such as an analogue open-reel tape-based master or some rare vinyl in need of remastering, the machines playing these need to be of the best quality possible. This audio will need to be digitized using your analog-to-digital convertor (ADC), ready for you to process the audio in your studio.

Capturing to the highest possible standards is expected when dealing with analogue sources. It is important to capture at the correct sample rate and bit depth for the clients' end needs. When capturing, it is sometimes sensible to use high sample rates to preempt any future audio developments. Data files will come into the studio in various forms, from differing file formats to very different sample rates and bit depths. The best approach would be to request that each source (studio) provides the files in the format that is best for you to work in, but sometimes this is not easy to achieve due to the equipment used in production. For example you might prefer 96 kHz, 24 bit when the studio can only supply 44.1 kHz, 24 bit. Therefore it will be necessary for you to adjust the files to fit with the native file format you'll be mastering with. We discuss these issues in the Knowledgebase in this chapter.

3.13 Processing

The next conceptual stage is processing. If the audio is coming from an external analogue source, as opposed to that of a data file, it is possible that the engineer might master the material on the way in, but it is more likely that the audio will just be captured and then the processing applied from the centrepiece of the DAW.

Processing is the mastering engineer's tool-box and will be brought out to repair, alter, direct and shape the audio material until the album, or track, is ready. A good analogy is a sculpture where the mastering engineer will chip away at controls until the flow and body of the album is complete and visible as a whole. Only at the end of the process will the sculpture come sharply into focus.

To appreciate processing in its mastering context, we've split the following chapters into the broad parameters we have to alter: dynamics, timbre, loudness, width and depth. Within these chapters we discuss processing in more detail as we come to understand some of the techniques available to us for manipulating mixes and producing refined masters.

Processing is the stage in mastering that many get excited about, and rightly so. However, it is not the be all or end all of the mastering engineer's role. There are important, less glamorous, aspects of the job that everyone, whether in the project studio or at the top mastering houses in the world need to engage in: sequencing and delivery.

3.14 Sequencing

As mentioned earlier in this chapter, achieving flow is, to us anyway, incredibly important when working in an album context. Even flow can be applied to single-track masters too, as this track will still need to flow with other songs on the listener's device.

Sequencing in mastering is the placement of one track against another in a track order. Sometimes this track order will be provided by your artist, or may be something they leave completely to you. In either case, you need to achieve the aforementioned flow. Sequencing includes things such as broad track levels, where despite much processing having taken place, we can still further adapt levels between tracks to ensure they seamlessly play together.

Also important are the gaps between each track. Choosing the 'right' gap will come down to experience and how the two tracks work well aligned together. We discuss more of the intricacies of this in Chapter 9. Allied to this, we also discuss fade ins and fade outs, how

Sample Rate Conversion (SRC)

As with a reduction in wordlength, it will also be necessary to reduce larger sample rates of say 96 Hz down to the necessary 44.1 kHz for inclusion on a CD. This is known as downsampling and is catered for digitally within all DAWs. There is less concern over the affect that sample rate conversion has over audio quality as dither has, and as such there are fewer editable controls. Whatever control you have over this, it is worth choosing the highest possible quality conversion process available within your DAW.

to manage them, and how to create playable music. For the mastering engineer, it is sometimes interesting to be asked to carry out segues between two tracks. This is where the tracks are somehow merged as one audio programme and 'glued' together, either by fading into each other, or by adding some kind of sound effect between them.

Most importantly, the sequencing stage gives the mastering engineer the ability to achieve a consistent tone across the album, track-by-track and as a whole. Despite each track having been processed, you may feel the need to make some final tweaks in response to the track order, especially if this has been changed during the mastering process, which can happen. All these aspects are covered in Chapter 9, after the chapters on processing.

3.15 Delivery

The final responsibilities for the mastering engineer are the less glamorous tasks associated with delivering a product. As time has gone on delivery has become more about uploading tracks to various servers – whether that is for a duplication plant using DDP, or whether they are single tracks destined for a music-sharing website.

In either case, it is important to understand the processes and tasks these both require of you, and to ensure that the relevant information is requested well in advance of delivery. In this part of the mastering process we're expected to not only deliver the music in its appropriate form, but also to take extreme care as we pull together ISRC and UPC/EAN codes, track titles, catalog numbers, CD-TEXT and many other aspects. More on all these later.

Getting any of the above wrong can play havoc with your finances, and only recently have insurers been in a position to mitigate against any problems in this area. Getting this right is of paramount importance and your artists and producers should play an active part in getting and checking this information is correct.

Once the details are in the system and the audio ready to go, it is important to prepare the product for its intended destination. If being pressed to compact disc, then it is highly recommended that a DDPi (Disc Description Protocol image) is produced and sent to the duplicators either as a DVD-R or more likely via the Internet. This is covered in more detail in Chapter 10.

CHAPTER 4

Controlling Dynamics

In this chapter

4.1 Introduction 58
4.2 Basics of compression 59
Threshold 59
Ratio 59
Attack and release 60
'Auto' settings 62
Knee 63
Gain makeup 64
Beyond the basics 64
4.3 Types of compressor 65
Optical 65
Variable-MU 66
FET 67
VCA 68
4.4 Compression techniques 69
Gentle mastering compression 69
Over-easy 70
Heavy compression 71
Peak slicing 72
Glue 74
Classical parallel compression 74
New York parallel compression 76
Multiple stages 76
4.5 Using side-chain filtering 78
4.6 From broadband to multiband 79
4.7 Setting up a multiband compressor 80
Step 1: Setting the crossover points and the amount of bands 80
Step 2: Apply compression on all bands to see what happens 81

Step 3: Controlling bass 82
Step 4: Controlling highs 83
Step 5: Controlling mids 84
Step 6: Gain makeup 84
4.8 Compression in the M/S dimension 85
Compressing the mid signal 86
Compressing the sides of the mix 86
4.9 Other dynamic tools – expansion 86
Downwards expansion 86
Upwards expansion 89

4.1 Introduction

Alongside pitch and rhythm, dynamics form one of the fundamental building blocks of music, helping define both the energy and emotional structure of a piece of music. Put simply, music would be considerably more lifeless without some sense of dynamic structure. However, although dynamics themselves are a good thing, it is important that a recording has some degree of dynamic control and refinement, so that the quiet passages are audible without having to turn your loudspeakers up to 11, and that the louder sections of the mix convey all the appropriate energy without making your ears bleed!

While it would nice to think that all finished mixes are presented with a polished dynamic structure, it's often the case that compression, in some shape or form, needs to be used to shape dynamic structure in some way. To shape dynamics, a mastering engineer can turn to a range of different tools: from analogue Variable-MU devices used from the earliest days of recording, right through the latest multi-band compressors that can re-shape a master in some radical ways. Understanding where and how to employ these processors, therefore, is essential in controlling dynamics in a way that is most empathetic to the music you're trying to process.

In this chapter, we're going to take a closer look at the techniques behind the control of dynamics and the principle tool used to do this – a compressor. We take a detailed look behind the parameters of compression and how these controls interact with each other to create some unique and interesting forms of dynamic manipulation. We'll also look at a range of specific techniques that allow compressors to be such a versatile audio tool: from subtle 'over-easy' techniques to parallel compression and audio 'glue'.

4.2 Basics of compression

As we saw briefly in Chapter 3, a compressor works as an automated gain control device, applying gain reduction in response to signal levels driving the input of the compressor. Put simply, a compressor attenuates loud signals, creating a more uniform dynamic, which in turn leads to an increase in the average signal level. A compressor is a useful tool for restricting the dynamic range of a master, although as we'll see, there are many other ways in which compressors can improve the sound of your finished product.

Despite the huge variety of compressor designs, the principle parameters of compression – threshold, ratio, attack and release – remain a consistent part of process. Understand these key parameters, and you'll be able to work with any compressor you're presented with, even if the controls seem a little unfamiliar at first.

Threshold

A compressor's *threshold* control governs the point at which gain reduction begins and forms one of the defining qualities of how the compressor behaves. Picking a low threshold, for example, will make the compressor work with more with of body of the signal, whereas a higher threshold makes the compressor more responsive to peak signals.

Ratio

The amount of *gain reduction* applied by the compressor as the signal exceeds the threshold is defined by the *ratio*. In effect, you can think of ratio as being the aggressiveness of the compressor – with softer ratios (such as 2:1, for example) creating subtle gain reduction movements, while harder ratios (such as 8:1 or more) brutally attenuate signals exceeding the threshold. Technically speaking, the ratio defines the increase in output in relation to an increase in the input. On a 2:1 ratio, for example, an input exceeding the threshold by 8 dB would be attenuated by 4 dB (or half as much, in other words). For every 2 dB increase in input, therefore, there's a 1 dB increase in the output level.

Of course, the amount of compression that you eventually achieve is a product of both threshold and ratio. For example, it's possible to use a hard ratio – say 8:1 or more – and yield just a small amount of gain reduction through the threshold being set particularly high. Likewise, low ratios could produce lots of gain reduction with the threshold being set at particularly low level. As you can see, compression is fundamentally about the interaction of these two parameters, and of course, the music you're trying to process, so avoid thinking that either control – threshold or ratio – is the only important part of the equation.

Threshold and Ratio are the two defining parameters of compression, governing when and how the compressor moves into gain reduction.

Attack and release

The attack and release times are often the trickiest part of compression to comprehend – both conceptually and (for the newcomer) audibly. However, following on from threshold and ratio, the attack and release settings define the character and effectiveness of the compression. Get attack and release right, and you'll deliver a tight, punchy sound. Set attack and release incorrectly, though, and your compressor will sound sloppy and ineffective, ruining the energy and natural dynamics of the music you are trying to process.

As you'd expect, the attack and release of the compressor govern the device's movements in and out of compression. Unlike a noise gate, though, the role of these controls is a little more counter intuitive than you might first imagine.

Let's start with the attack of the compressor, which is the easiest component to hear. Technically speaking, attack sets the speed at which the compressor starts to apply gain reduction over and above the threshold. A good way of understanding the principle is to imagine two different audio engineers working as 'human' compressors 'gain riding' a physical fader. The two engineers have different reaction times – one is quick off the mark, alert and responsive, while the other is a somewhat tired and jaded, always slightly behind the levels that are presented to them.

In effect, the example here illustrates both a fast attack time (from the responsive engineer) and a slow attack time (from the engineer that's slower off-the-mark). What's interesting is to start making assessments of their relative performance. At first, you can imagine the fast-acting engineer as being the best. Certainly, they're effective at their task – grabbing peaks quickly and applying plenty of gain reduction. However, in their haste they start compressing important transient detail, arguably drawing some of the life out from the sound. In some respects, maybe the more relaxed approach has its merits!

What this example illustrates are the pros and cons of different attack settings and why differentiation is so important. Contrary to what you might expect, a slow attack actually preserves the attack – or transient detail – in the signal you're trying to process. However, a fast attack is good at actually pinning-down transients, almost softening the edges slightly.

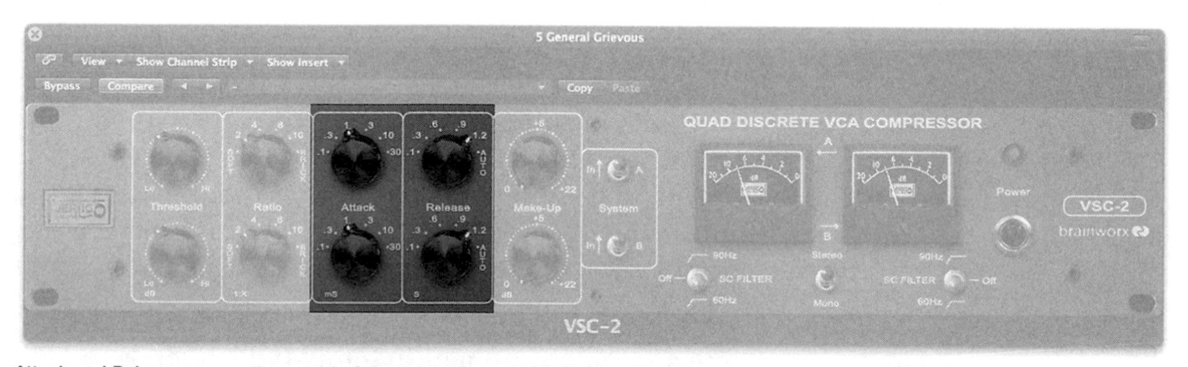

Attack and Release govern the speed of movement in and out of gain reduction.

Following the logic of the attack section, release sets the speed that a compressor returns to its neutral state after the application of gain reduction. Again, you'd perhaps think that a fast-acting release time is best, so that the compressor isn't spending any more time compressing than it needs to be. However, having such a fast release setting doesn't make the compressor particularly subtle, as the 3–6 dB it attenuated a few seconds ago is suddenly pumped back into the output. In short, we hear the compressor going about its business, which isn't a good thing!

In reality, it's best that the release time matches the natural envelope of the music you're trying to process. Going back to the engineer analogy, it's almost as if the less responsive engineer has an empathy and respect for the music, bringing the fader up slowly enough so that you don't notice what they're trying to achieve. By only being interested in the absolute signal levels, however, the other 'responsive' engineer's input somewhat compromises the effectiveness of the music, although, of course, they've been very efficient at their particular job!

Two important qualities of compression – that of a compressor 'breathing' or 'pumping' – can be directly attributed to attack and release settings, forming a big part of the character of a compressor. As we'll see later on, a large part of the sonic character between different designs of compressor can also be directly attributed to their particular attack and release characteristics, with some designs being more responsive than others. Understanding how a compressor 'moves', therefore, is vital not only to controlling dynamics, but also to adding character and interest into your masters.

'Auto' settings

For the uninitiated, the inclusion of an *auto* setting in relation to the attack or release times could appear to be a form of 'idiot button' – a tool that negates you paying any interest in respect to the exact setup of the unit. In mastering, however, the use of an auto release setting could have some merit, and is well worth considering as a default setting, or in other words, as starting point for further exploration.

Unlike processing a single instrument in a mix (which tends to have uniform release characteristics defined by the instrument you're processing) the release characteristics of a full mix are complex and ever changing, formed by the a layered set of contrasting instruments. Put simply, it's often hard, if not impossible, to find an empathetic release curve that matches the shape of the music. An auto setting, however, especially if it is part of a mastering-grade 'buss compressor', is programme dependent – in other words, it adapts itself based on the material you're trying to process.

An auto release setting might feel like an 'idiot' button, but it performs an important role given its adaptive properties.

If you want a transparent-sounding compression, therefore, an auto release setting is often a logical choice, even if you're clear about how a compressor's release parameter behaves. Disengaging the auto release setting will, of course, deliver more character to the result and allow you to specify exactly how the compressor behaves, but it's sonic footprint might become a little more apparent!

Knee

The *knee* of a compressor defines the transition between a linear response to gain (before the compressors starts applying gain reduction), and a non-linear response to gain (where the gain reduction circuitry becomes active). In theory, the difference between these

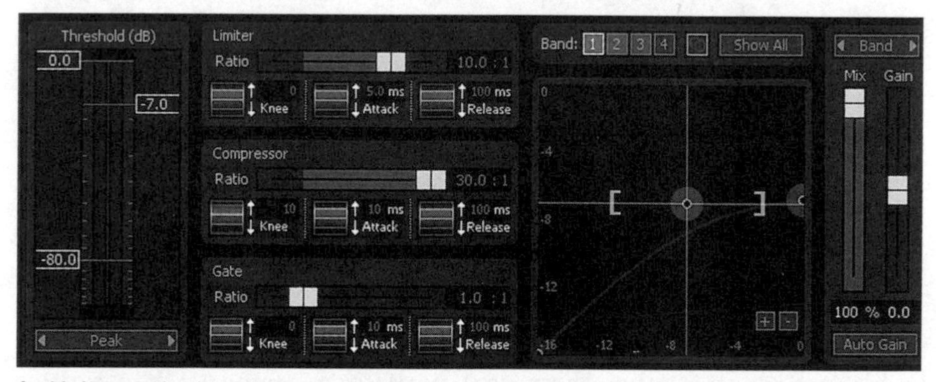

A wide knee setting (as shown on this Input/Output graph on Ozone) creates a smooth transition into gain reduction.

two states can make compression more audible, with a listener more aware of when the compressor is and isn't working. With a softer knee setting, the compressor starts applying a small amount of gain reduction ahead the threshold, only hitting its full ratio over and above the threshold. In effect, a knee creates a form of variable ratio, that increases in response to input levels.

Gain makeup

Compression is, of course, a process of attenuation – turning down the amplitude in response to signals rising above the threshold. Therefore, to preserve appropriate gain structure, as well as to appreciate the positive sonic effects of compression, we need to increase the output gain using a control called *gain makeup*.

As a basic guide, the amount of gain makeup should be broadly similar to the amount of gain reduction that's been applied, less a few decibels or so. So, in the case of a compressor applying 6 dB of gain reduction, you can raise the gain makeup by around 4 dB. Of course, variables such as the attack and release settings and any peak signals still present in the output will ultimately dictate how much gain makeup you can apply – so pay close attention to the output meters and ensure you leave some headroom for any subsequent devices in the signal path.

As compression creates an overall loss in level, we need to consider raising the output gain using a Makeup control.

Beyond the basics

Beyond these basic controls, you'll find a host of other options on a variety of compressors, including side chains, key filters, mix controls, and so on. Rather than deal with them here, we'll introduce these additional concepts as we move through some of the more advanced topics covered later on in the chapter.

4.3 Types of compressor

Having understood the key parameters of compression, lets take a look at some of the different designs of compressor that are available. Distinguishing between these different designs – either in software or hardware – is essential in picking the right tool for the job. It also highlights the historical evolution for compressors, and how technical innovation might have ushered in some significant extra flexibilities, but also occasionally losses in respect to character and sonic interest.

Later on, as we move through different styles of compression, we'll also highlight how specific designs of compressor are better at some tasks than others.

Optical

Example: Teletronic LA-2 A, Shadow Hills Mastering Compressor

By comparison to today's digital signal processing, an *optical* compressor is a relatively crude form of signal processor, but also one with a surprising degree of musicality. As the name suggest, the input of an optical compressor is fed to a light bulb or LED, which glows brighter or darker in response to the incoming signal levels. In turn, the strength of the light is read by a photocell, which is then used to control the amount of gain reduction applied. Put simply, as the light glows brighter, a greater amount of compression is applied.

What's so interesting about an optical compressor is its unique response times in respect to the photocell gain control element. In effect, the light has an inherent lag to it – being sluggish in its response to transients, and slow to fully dissipate its energy after long periods of high-energy signal. As a result, optical compressors have a unique slow-acting attack time, while the release almost has a two-stage quality to it, being initially quick, but then slow to

Optical compressors are often slow to react, but they provide a surprisingly musical form of gain reduction.

completely return to its null state. Although technically inefficient, the response of an optical compressor is incredibly 'musical' and sympathetic to many of the sources you pass through it.

While older-style optical compressors see plenty of use in tracking and mixing, their application in mastering needs a degree of caution, as there's often no total control over the attack and release settings. However, there are some examples of hardware master compressors featuring optical stages, such as the Shadow Hills Mastering Compressor, and of course PrismSound's Maselec Master Series MLA-2 Precision Stereo Compressor is based on an optical gain cell.

Variable-MU

Examples: Fairchild 670, Manley Variable Mu Limiter/Compressor

Variable-MU compression is another older design of gain control, this time using a re-biased vacuum tube as the gain control element. What's interesting about a variable-MU compressor, particularly to

Variable-MU compressors don't offer the fastest transient response, but they compress in a musical way.

the application of mastering, is the lack of a traditional ratio control. In effect, a variable-MU compressor could be compared to an ultra-wide soft knee, with the ratio effectively becoming stronger as the compressor is pushed harder. Rather than working with threshold and ratio, an engineer needs to finely tune the compressor's input and output controls – either driving the compressor hard for a more extreme compression, or in a more reserved setting if you want the compression to be subtler.

Although attack and release controls are provided, it would be fair to say that a variable-MU compressor isn't the fastest-acting compressor in the world. As such, you'll probably want to avoid using variable-MU designs for any kind of aggressive peak limiting activities. Instead, use a variable-MU compressor to massage the body of the track into place, often using deliberately graduated attack and release settings so that the compression is smooth. Of course, as the compression is tube-based, it's also likely that the variable-MU design will offer a touch of extra colour to your master, which may be as beneficial as the compression itself!

FET

Example: *Universal Audio 2-1176*

FET compressors, which use a 'field effect transistor' for their gain control, where originally developed as a snappier alternative to variable-MU and optical designs. Even to this day, FET compressors are famed for their ultra-fast attack and release settings, making them a much more suitable choices for peak-style compression, and for more extreme 'colorful' compression.

Despite having superior attack and release times, FET compressors are often more coloured than other types of compression, and aren't

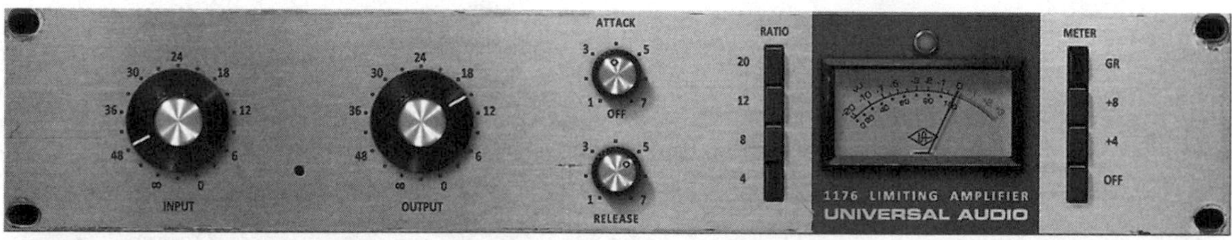

FET compressors are fast-acting with plenty of character. Not the subtlest compression in the box, but they have a great colour to add in the right situations.

widely used in mastering. That said, if you're looking at techniques such as parallel compression (which we're taking a closer look at later on in this chapter), the distinctive fast-acting and colourful sound of an FET compressor may well be an interesting avenue worth exploring. However, if you're look for smooth transparent compression, you're in the wrong place!

VCA

Examples: *Neve 33609, SSL XLogic G-series, Vertigo Sound VSC-2.*

In many ways, *VCA* compression represents the principle of technical and sonic development in respect to the design of a compressor. By using a 'voltage-controlled amplifier' for its gain control, a VCA compressor can provide plenty of smooth gain reduction and full control over its attack and release settings. In short, VCA compressors are versatile – adaptable enough to provide smooth and sumptuous 'glue' to a master, or aggressive enough to clamp down on transients. For a mastering engineer, this versatility is an important quality, especially if they're working with a huge variety of musical styles, each needing a slightly different approach to dynamic control.

However, before we crown the VCA as the king of compressors, it's worth pointing out that there are some huge variations between the quality and performance of different VCA designs. Given the relative cost efficiency of producing a VCA compressor in comparison to optical, FET or variable-MU designs (which all require expensive discrete components), a lot of VCA compressors have been built to provide 'serviceable' audio performance, rather than audio excellence. Rather than using any VCA compressor, look at designs that are specifically aimed at mix-buss processing, such as the Neve 33609, SSL XLogic G-series, or the Vertigo Sound VSC-2.

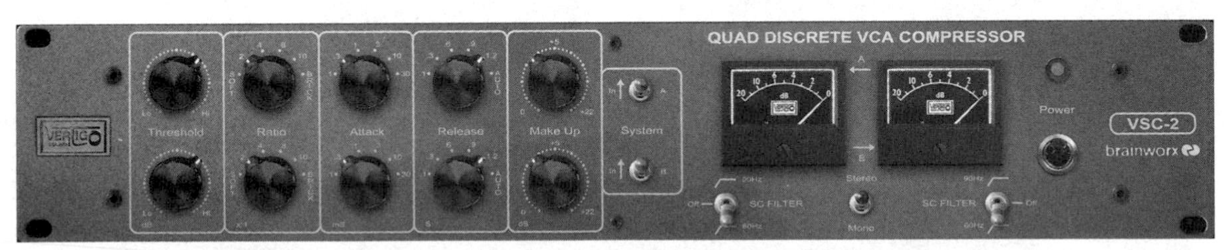

A good mastering-grade VCA compressor is one of the most flexible and dependable tools to have at your disposal.

4.4 Compression techniques

Having looked at the principles of compression and the variety of compression designs that are available, let's now explore the palette of basic compression settings and techniques a mastering engineer will use. Each setting needs tweaking for the type of music you're working on, but it provides a suitable starting point for further exploration.

Although we've provided some typical parameter values, some settings are deliberately vague (particularly threshold) and they shouldn't be thought of as absolute values, but starting points for you own experiments. What's more important is the intention behind the settings, which will inform your efforts to tweak them in response to the unique attributes of the music you're trying to process.

Gentle mastering compression

Ratio: *1.5:1 (or less)*
Threshold: *Low*
Attack: *30 ms (medium – slow)*
Release: *300 ms (medium) or Auto*

This first setting is the defining starting point for mastering compression, and a technique that provides a pleasing dynamic control to almost any music you pass through it. The two crucial ingredients, of course, are a low ratio (around 1.5:1 through to 2.5:1) and a low threshold. In theory, the compressor spends most of its time in gain reduction, only ever applying just a few decibels of dynamic control, and aiming for around 3 dB on your gain reduction meters. Keep the attack and release on a medium setting, ensuring that the transients aren't compromised too much, and that the release is slow enough to be empathetic to the envelope of the music (indeed, an auto release might work well here).

Overall, this low-ratio compression should change how the 'body' of the track behaves, reigning in some of the excesses of the chorus, and bringing up some of the energy in the verse due to the lighter amount of gain reduction being applied. What shouldn't be too apparent, though, is the compressor going about its business, particularly as there's no sharp transition between the compressor going in and out of gain reduction. Of course, you won't achieve a huge amount of reduction in the dynamic range, but your master will still sound musical and have room to breathe.

Using a low ratio and threshold setting allows you to control the 'body' of the track in an effective way.

Over-easy

Ratio: 2:1
Threshold: Low to medium
Knee: Soft (or use a compressor with a wide knee)
Attack: 10–50 ms (medium)
Release: 300 ms/Auto

In essence, the over-easy technique is an adaptation of the basic gentle mastering compression we looked at in the last example. The trick here, though, is to make use of either a compressor with a noted 'wide knee' response, or a compressor with a variable-knee setting. Good examples here would be a variable-MU compressor, which has a continuously variable ratio based on how hard the compressor is driven, and a good VCA buss compressor such as the SSL-Series G, which has a unique super-wide knee on its 2:1 ratio.

Configuring the exact settings will vary given the compressor you're working on and the music you're trying to process, but you should aim for the compressor spending plenty of time in gain reduction – applying small amounts of dynamic control and only hitting about 3–4 dB of gain reduction when the music has some real energy behind it. The combination of the wide knee and/or low threshold should be what keeps the compressor working, arguably coupled with empathetic attack and release settings to keep the movement transparent and subtle.

Sonically speaking, over-easy compression should sound similar to our initial general mastering compression setting we explored in the first example. This isn't a style that delivers lots of reduction in dynamic range, nor should it be too noticeable. They key point, however, is that you start to create some distinction between a lighter ratio in the verse, and a pushier ratio in the chorus. Rather than a one-size-fits-all 'broad brush' mastering compression, you're starting to create a little more distinction between the different passages in the music.

Over-easy compression makes a defining use of a wide knee setting. Try using a Variable-MU compressor, or something like the SSL Buss Compressor, which has a wide knee on its 2:1 ratio.

Heavy compression

Ratio: 4:1
Threshold: Medium to high
Attack: 10 ms (fast)
Release: 100–300 ms (fine-tune to track's tempo)

Once we break the ceiling of around 2.5:1, our compression should start to sound much more noticeable. Put simply, we're actively using the 'sound' of compression in our master, rather than just gently controlling the dynamic range. While heavy handed compression might not be appropriate on subtle acoustic music, it can give more energetic tracks a distinct lift and some important additional intensity.

If you're wanting a touch more 'sound' to your compression, we'd argue that a 'magic' starting point is a 4:1 ratio – a setting that's distinctly harder than what we've looked at so far. As well as raising the ratio, we also need to raise the threshold, giving more focus to the compressor going in and out of gain reduction – in short, we need to distinguish between passages the compressor can largely ignore and passages it needs to work on. Unlike mixing, you maybe don't want to start exploring extreme amounts of gain reduction, so try to set the threshold so that the compressor is applying between 2 and 4 dB of gain reduction.

The finesse of heavier compression comes with the attack and release settings, especially as this forms the principle sound of the compressor going about its business. Start from a suitable 'vanilla' setting – with attack around 10 ms, and release around 100 ms. Lengthening the attack time will let more of the transient energy through from the track, which can help add percussive bite, but might also impinge on the compressor's ability to control peak signals.

The release time is important to get the compressor breathing in the correct way. Ideally, the release shouldn't be too fast, but graduated to match the feel and tempo of the track. As a rough guide, if the compressor can't restore most of its gain between beats then the setting is probably too slow. However, if the release is too fast, you might not experience the pleasant 'massaging' effect the compression can deliver.

Using a heavy compression setting will deliver more energy to the track, although you'll need to ensure that the effect doesn't become too 'pumpy'.

Peak slicing

Ratio: 8:1 up to inf:1
Threshold: High

Attack: *Fast*
Release: *Fast*

Restricting the peak transients can be an important way of controlling the overall energy of a track, particularly in examples of music with a lot of varying transient levels. Ideally, though, this type of compression should be a subtle dynamic control rather than a form of deliberate signal abuse to create extreme loudness (something we'll explore in Chapter 6). Think of this peak slicing as the conceptual opposite to 'gentle mastering compression', although both should be equally transparent.

The key to effective peak slicing begins with a high threshold and ratio setting. Unlike gentle mastering compression, the compressor should only be called into action sporadically, kissing about 3 dB off the loudest parts of the mix. For a transparent result, you'll also want a fast attack and release setting, which definitely rules out the application of optical or variable-MU compressors in this role.

Distinction should be made between the techniques of brick-wall limiting, which we'll explore in greater detail in Chapter 6, and this musical form of peak control usually applied by a VCA compressor. In short, your objective here isn't to push a track right up to 0dBFS, but instead to provide a little more control to the peaks as a means of letting the rest of the track articulate itself in an effective way. In truth, some transient energy will probably still get through, but the overall dynamic should seem proportionately more controlled and defined.

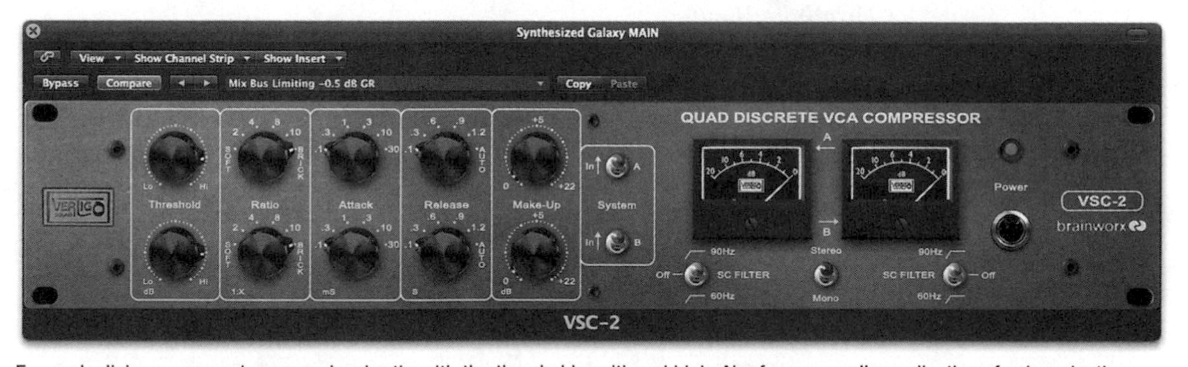

For peak slicing compression, use a hard ratio with the threshold positioned high. Aim for a sporadic application of gain reduction.

Glue

Settings: Variable!

In truth, the previous four examples (gentle mastering compression, over-easy, heavy compression, and peak slicing) could all be used as forms of 'glue': using a compressor to help a mix gel together. In reality, therefore, the entity of glue is just a by-product of gain control, making the track sound like a whole entity rather than its individual parts. From a technical perspective, this could be explained by the fact that all the instruments in the mix are being subjected to the same gain reduction – in effect, the movements of the compressor in and out of gain reduction gives the instruments a common identity.

However, discerning how and why a compressor can glue a mix together is an important part of understanding mastering. Put simply, it's often the case that a compressor is used as much for its glue-like tendencies as it is to control the dynamic range of the input, or indeed, to add loudness. Rather than just looking at the gain reduction meters and the overall reduction in dynamic range, listen carefully to how the compression changes the identity and cohesiveness of the track. By adapting the type of compressor you use, or the style of compression, you might achieve little difference with regard to the change of dynamic range, but have a more significant impact on how the mix is glued together.

In conclusion, glue is less about a specific 'magic' setting but more about a way of listening and critically analyzing the results of what you achieve using a compressor.

Classical parallel compression

Ratio: 2:1
Threshold: Low
Attack: Medium
Release: Medium
Mix: 50 per cent

Parallel compression is a real buzzword in audio production circles, and a technique that's well worth closer inspection. It's also interesting to note its two contrasting applications, first as a refined tool for classical mastering, and secondly as a 'mojo' enhancing tool for popular music.

In its original form, parallel compression was developed in classical recording circles as an alternative means of applying compression.

What these engineers were trying to achieve was a compression that, to some extent, reversed the usual logic of applying gain reduction.

As we've seen from our previous examples of compression, gain control is achieved by turning down signals above a threshold – the louder, more exciting parts of the track are squashed, while the quieter segments pass unaffected. But what if you could approach gain control in a different way, turning up quiet sounds rather than turning down loud ones?

The solution, in this example, is to patch the compressor in a different way: splitting the signal into two parallel paths – one with and one without compression – with the resulting effect formed from a blend of the compressed and uncompressed outputs. Using a low threshold setting of around −50 dB (the compressor tends to stay in gain reduction for most of the time) and a soft ratio (2:1 or below), the compressor is configured to contribute most to the overall output during the quieter passages of music. As the track gets louder, gain reduction starts to be applied even more substantially, to a point at which the compressor's contribution to the overall programme level is negligible.

Compression applied in this way produces a far more transparent and musical effect than traditional compression, although it does tend to assume that the music has a wide dynamic range for the compression

Here's the routing for a classic parallel compression effect. A 'hard acting' compression is patched across an insert, with the amount of compression blended-in with the main signal.

to work with (which is why it sounds so effective on classical music). Transients are preserved, less distortion is evident, and the amount of boost to quieter sections can be easily controlled by feeding more or less of the compressed signal into the final output.

New York parallel compression

Ratio: *4:1 up to inf:1*
Threshold: *Medium to high*
Attack: *10 ms (fast)*
Release: *Adjust to track's tempo – around 200–300 ms*
Mix: *50 per cent*

Here's a slightly different twist on the parallel compression technique, which demonstrates a completely different application of the theory. Again the overarching concept is a 'blended' compression, having both a compressed and uncompressed version as part of the same output. The aim is to add the intensity of compression, but without any damage to the transient energy in the mix.

Start by configuring your compressed channel. Here, it's important that we're not too subtle, and that we're really attacking the transients in the mix. A good setting, therefore, would be a 4:1 ratio, with a fast attack and graduated/medium release. You should aim for some deliberate transient reduction, a good dose of additional level, and a release time that restores itself across the beat. Don't be afraid if you're pushing 6 dB of gain reduction, as long as the compressor is doing some work! In more extreme examples, it has been known to EQ the compressed signal, although you need to keep a close eye on any phase issues that might develop as a result of this.

To create the finished result, blend the transient-reduced compression alongside the uncompressed version. Given that the compressed channel has most of its transient energy removed, you won't add a significant amount to the peak signals in the output, but you will add to the overall body of the track. Changing the balance between the two channels, therefore, is much the same as setting a wet/dry mix on the output of a reverb, although in this case we're defining the amount of compressed signal returned into the mix

Multiple stages

First stage
Ratio: *2:1*

Threshold: *Low*
Attack: *Medium*
Release: *Medium*

Second stage
Ratio: *8:1*
Threshold: *High*
Attack: *Fast*
Release: *Auto*

So far, we've looked at single stages of compression – that is, using a single compressor to achieve all your desired dynamic control. However, there's no reason to think that a single compressor is the

Contrasting two stages of compression is a great way creating a balanced overall dynamic. Start with a softer 'massaging' compression, then add a peak-limiting stage as the next device in the signal path.

one and only tool you can use to shape your dynamics. It's often the case that different techniques and designs of compressors are applied in series to create the desired end result.

As always, the key to multiple stages of compression is to have some defined strategy and objective in mind. In the 'good old days' of recording, engineers often had very little equipment at their disposal, so, as a result, the application tended to be focussed and appropriate to the objectives in mind. In the software world, however, it's tempting to add another plug-in to solve a problem, rather than look at your existing settings. Remember, less is often more, but in some exceptional cases two compressors might be better than one!

If you do use two stages of compression, make sure each stage is contributing in a specific way. For example, you might want to contrast a light variable-MU 'over-easy' compression with a more transient-squashing VCA compressor. Equally, you could blend a small amount of parallel compression to add body to a mix that has been compressed lightly with a 1.5:1 ratio. In both of these cases, each compressor brings something different, making the overall effect proportionately more interesting.

4.5 Using side-chain filtering

One of the bugbears of any buss compression is the input the bass spectrum has on the amount of gain reduction applied. As a large amount of sound energy is contained in the bottom-end of the mix, it can often have a disproportionate input on the amount of gain reduction applied, even though our ears are less sensitive to that part

Side chain filters can be used to control excessive amount of gain reduction triggered by the low-end of the mix.

of the spectrum. Put simply, the kick drum and bass often dictate how much compression is applied, rather than other equally important components such as the snare or lead vocal. As a result, the high-end of the mix – particularly cymbals – often seems to pump up and down in an unnatural way, often at odds with the dynamics in that part of the sound spectrum.

By using a side-chain filter, however, it's possible to differentiate between the signal that the compressor is listening to and the signal we want to be compressed. By attenuating the bass end of the signal feeding the side-chain input (usually using a simple bass roll-off control) we can make the compressor less sensitive to the bass components of the sound spectrum. With the side-chain filter applied, the compression is much more in line with how our ears respond to sound, and as a result it seems more natural to our ears. Of course, the downside is that the amount of dynamic control you'll be able to apply might be slightly reduced, especially if you have problems in the bass-end of your mix. In these situations, a dose of multiband compression (explored in the next section) might be a better solution.

4.6 From broadband to multiband

So far our exploration of dynamic control has focussed on what could be termed 'broadband' compression – in other words, compression that is applied across the entirety of the frequency spectrum. With a well-balanced source, broadband compression can deliver great results, but as we've seen in the last section about side-chain filtering, issues in the bass spectrum can create unnecessary modulations in the upper part of the audio spectrum.

Multiband compression, therefore, is a useful alternative to broadband compression in situations where the mastering engineer needs more detailed control over by the dynamic and spectral properties of the track. By dividing an input into three or more discrete frequency bands, a multiband compressor can be far more 'directed' in respect to its application of compression, allowing each frequency band to be compressed in a different way, and also avoiding unwanted modulation and distortion created by one part of the frequency spectrum having a heavy-handed input on the amount of gain reduction applied.

Although multiband compression seems like a logical improvement over broadband compression, there are some important caveats worth bearing in mind in respect to its application. Unless you're very careful, it's highly likely that the multiband compressor will affect both the dynamic and timbral properties of your track, so be wary of multiband processing if the timbre feels well-balanced in the mix that's presented to you. Also, given how a multiband compressor slices up a track into three or more frequency bands, you need to pay close attention to any damage inflicted on instruments either side of a crossover point. In short, multiband compression is a dangerous tool in the wrong hands, but also a powerful and sonically effective tool when applied in a selective and considered way.

4.7 Setting up a multiband compressor

Given the amount of parameters involved with a multiband compressor, it's easy to become overwhelmed by its setup and operation. Whereas a broadband compressor usually presents no more than around six parameters to work with, a typical multiband compressor can often hit 30 or more parameters across three or more bands, and even more options if you also consider multiband expansion and limiting. As always, it pays to be clear from the start about what you want to achieve, and then go about exploring the controls until you hear something along those lines. Always check the bypassed version, applying a degree of level correction so that the extra level the compressor delivers doesn't fool you.

Step 1: Setting the crossover points and the amount of bands

Setting the crossover points between the three or so frequency bands might seem a straightforward task, but in reality it has a significant impact on all the actions you carry out later on in the process. Firstly, you need consider the amount of bands you want to use for compression. Although the compressor might feature up to six bands of operation, you should always veer for the 'less-is-more' ethos, using just enough bands to divide the mix as you feel

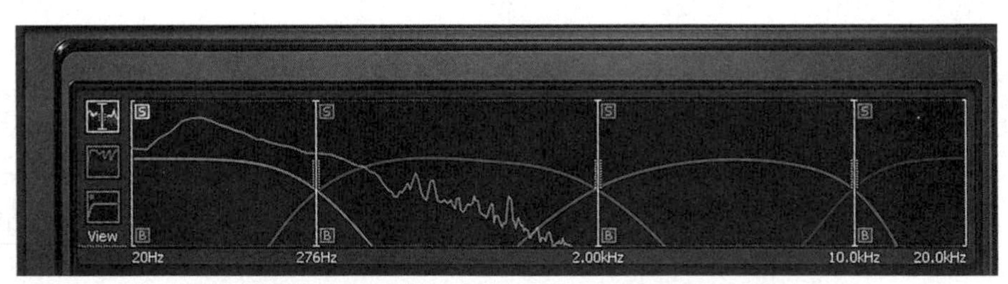

Position crossover points so that they divide your track in a musically appropriate way.

appropriate. More bands simply leads to more confusion, so it often pays dividends to stick with around three bands, dividing the track up to lows, mids and highs.

As a starting point for a simple three-band setup, consider placing the two crossovers at around 300 Hz and 4 kHz. If possible, listen to each band in isolation, checking that the key instrumentation (e.g. bass, vocals and lead guitar) sits happily in the three frequency areas you've defined. It might also be that you're attempting to control a specific frequency area of the mix (such as how the kick moves in the subsonic spectrum), in which case set the crossover points to meet your specific objective.

Step 2: Apply compression on all bands to see what happens

The next stage is to see how the multiband compression is behaving, and in that respect it's best to get all the bands 'moving' to

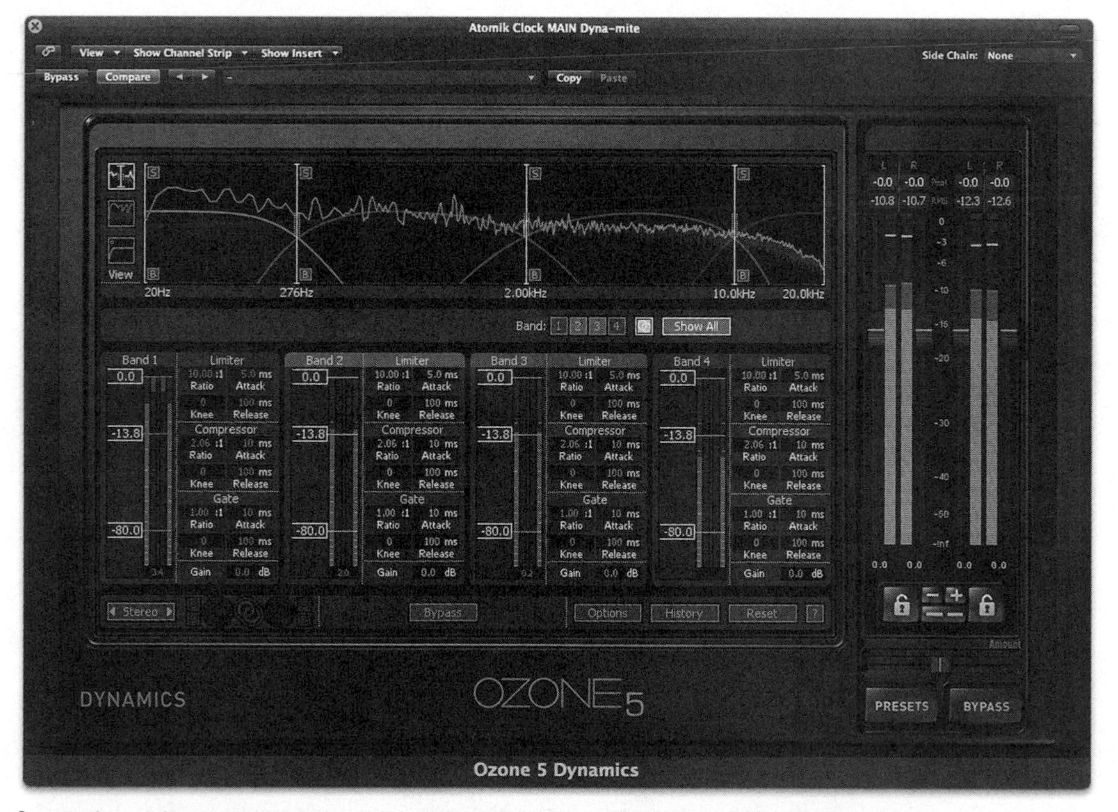

Once you've set the crossover points, see how the multiband compressor behaves with a 'vanilla' compressor setting.

some degree. If your compressor supports linked editing, try linking all the controls so that you only need to adjust one band's settings. Otherwise, set up the compressor using some of the light settings that we explored earlier on – setting a ratio between 1.4:1 and 2:1, a relatively low threshold, and an attack and release on medium settings (20 ms attack, 200 ms release).

Now listen to the results, and see how the compressor is behaving across the three bands. Note how different sounds are triggering their respective bands, whether it's a kick drum pulling on the bass-end, or some loud cymbals splashed on the highs. If any of the bands aren't working appropriately (maybe the bass-end is working too hard, for example, and no compression is being applied to the highs) considered adjusting the individual threshold controls (although of course, you'll need to remember to unlink the controls at this point).

Step 3: Controlling bass

With our general compression settings applied, we can now step through on a band-by-band basis and refine the settings in line with what we want to achieve. Each part of the frequency spectrum contains some interesting caveats and qualities we'll want to explore, so now it's important to start differentiating our approach to each band.

The bass part of the spectrum is a golden opportunity for the multiband compressor, as we start to define a solid foundation for the rest of the track to sit on. Of all the bands we have to play with, the bass spectrum arguably benefits from the most compression, helping to keep the bass controlled and the overall dynamic more even. Where the mix is particularly deficient in respect to the control in the bass (maybe the kick drum or bass guitar hasn't been compressed in the mix, for example) then a good dose of multiband compression can deliver some significant improvements. When you need a touch more control in the bass end, consider either lowering the threshold so that the compressor massages more of the sound, or setting a lower ratio so that the compressor works harder.

When it comes to refining the attack and release settings, keep a close eye on modulation distortion, whereby the compressor starts to work with the individual wavecycles rather than the notes themselves. This unwanted modulation distortion can be easily rectified by keeping the attack and release suitably smooth, rather than being too fast-acting.

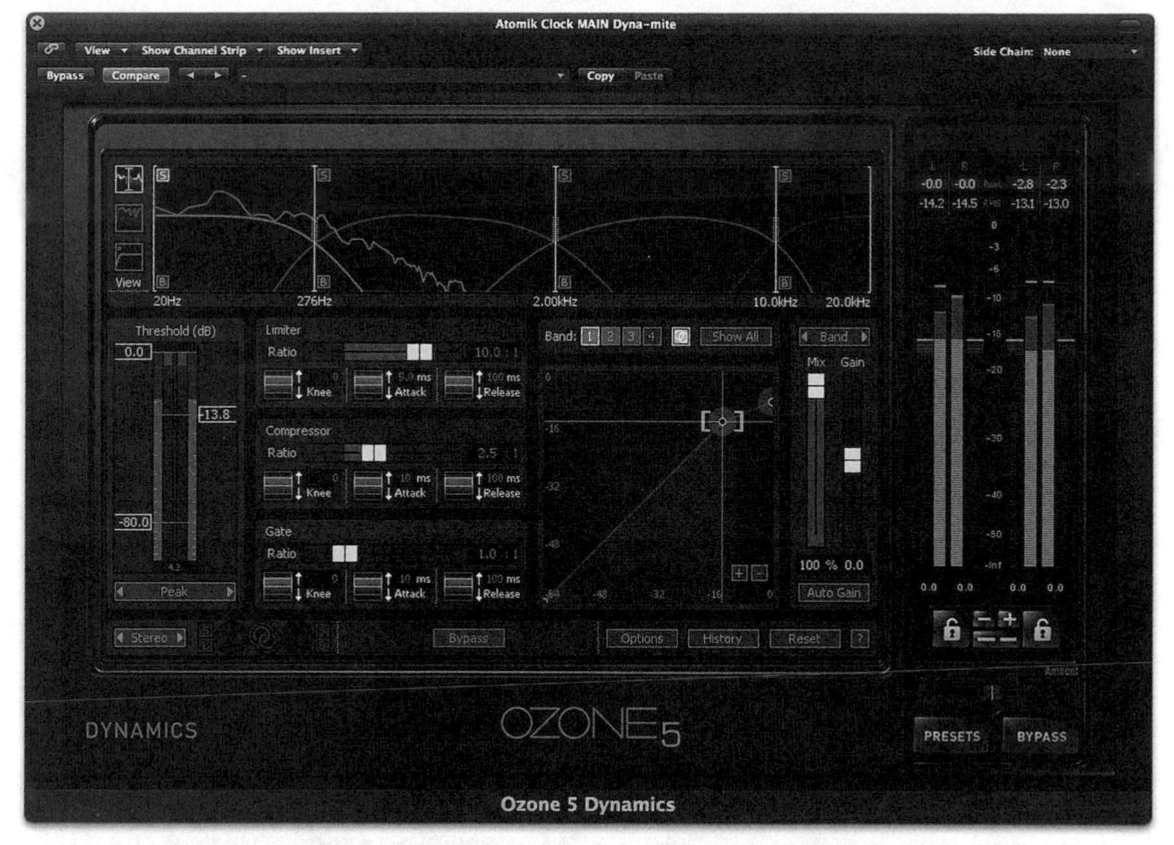

The bass-end of a mix is one of the most important areas for a multiband compressor, and an opportunity to really define the 'foundation' of your master.

Step 4: Controlling highs

At the other end of the spectrum we have the highs, which contain much of the all-important transient detail and sparkle of our master. As a general guide, you'll probably want to avoid over-compressing this area, mainly as the ear is particularly sensitive to excessive amplitude modulation in this part of the frequency spectrum. Instead, take a light approach and avoid too much gain reduction.

When it comes to the transient detail, you have a number of options. Firstly, increasing the attack time will help transient details remain intact, which might be important if you're over-compressing some of the other parts of the frequency spectrum. If you need a touch more transient control though, consider raising the threshold and ratio on the high band, as well as adjusting the attack to a relatively fast setting. Once set, the compressor working on the high band should only respond to the occasional transient, although care should be taken not to overcook this compression so as to remove all the energy and vitality from the track.

Step 5: Controlling mids

As you'd expect, the mid-range forms a conceptual counterbalance between the more heavy-handed compression applied to the low-end of the mix, and the softer settings applied to the highest band. In some respects, the default 'vanilla' setting should be more than appropriate here, maybe lowering the threshold if you want a touch more compression. Importantly, this frequency band contains a large amount of musical information – including lead vocals, piano and guitar – so touch of extra compression can help push this part of the mix forward.

Step 6: Gain makeup

In contrast to broadband compression, the gain makeup section of a multiband compressor is a significant part of its creative control. Put bluntly, the gain makeup of a multiband compressor is as much

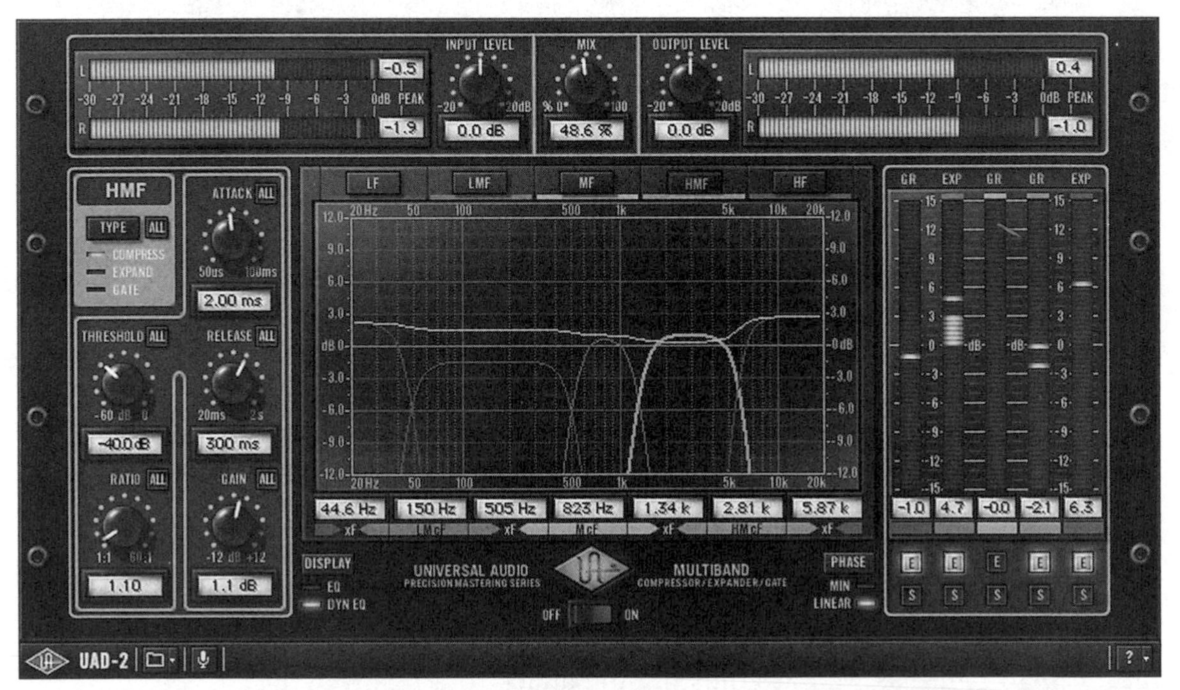

The gain makeup of a multiband compressor is a powerful tool, shaping both dynamic and timbral properties of the final master.

a tool for timbral modification as it is for dynamic control, a theme that we pick up in Chapter 5. You could, for example, simply choose to change the relative level of the bands by a few decibels – perhaps pushing the low-end slightly, or just notching-out some mids. Of course, adding differing amounts of compression changes the innate balance of the bands, so that as a band is compressed harder it's pushed towards the front of the mix.

Ultimately, care needs to be taken to ensure you don't sacrifice any of the innate musicality of the original tracks. Listen to any commercial radio station and you'll hear this effect in action, where the strength of a 'generic' multiband compressor strapped across the station's output creates some rather extreme modifications to the musicality of the majority of music that passes through it. If you're after some general dynamic control, therefore, broadband is best, and only turn to multiband processing in situations where you actively need the benefits it offers.

4.8 Compression in the M/S dimension

Compressing in the M/S domain is one of the more unique and interesting tools a mastering engineer has at their disposal. Of course, simply processing in M/S won't make your master sound good, so you need to have a clear objective as to why you're using this particular technique. In most cases, it comes down to the need to make some distinction between the mid and side channels of a mix – applying compression more to the mid channel, for example, rather than to the side channel. As you'd expect, this M/S compression will affect both the stereo qualities of the mix as well as its dynamics, just as a multiband compressor affects both timbre and dynamics.

As a point of reference, M/S compression was and still is an important part of mastering for vinyl, which is why compressors such as the Fairchild 670 have an M/S mode labelled as lateral/vertical compression. Vinyl is effectively cut in M/S with lateral movements of the needle forming the summed mid signal, with the vertical movements carrying the side. By compressing the lateral and vertical signals, a cutting engineer could control the movements of the needle, optimizing the cut and ensuring that the final record would playback effectively.

To understand the effect of M/S compression let's compare and contrast two applications – one with a bias towards the mid channel, the other with a bias towards the side.

Compressing the mid signal

Given that the principle instruments usually sit in the centre of the mix, any compression applied to the mid channel will change their presentation in the final mix, arguably making the mid channel more prominent and the stereo width slightly reduced. Used with care, mid channel compression is a great way of pushing the vocal forward in situations where the mix has been presented slightly vocal-light. Of course, you also need to consider the effect of mid-channel compression on other instruments, particularly the bass, although it is always possible to combine this M/S compression boost with an M/S equalization cut to remedy the problem.

Compressing the sides of the mix

Compressing the side channel is an interesting way of lifting the sides of a mix, increasing the width of the recording but also allowing some of the details away from the principle instrumentation to be enhanced – whether it's the return for a reverb, for example, or some backing vocals panned to 9 and 3 o'clock. Even subtle amounts of side compression can work well, adding a touch of extra life into a recording, especially if there's an existing bias towards instruments panned to the centre of the mix.

4.9 Other dynamic tools – expansion

So far, our exploration of dynamic control has largely focussed on compression – in other words, the application of gain reduction above a given threshold. However, this isn't the only way of applying gain. For example, what would be the result if gain reduction were applied to a signal below a given threshold? What we've created here is the polar opposite of compression – *expansion*.

Downwards expansion

As well see in this section, there're two forms of expansion that can be applied – *downwards expansion* and *upwards expansion*. Although the means of applying either form of expansion is the same, the

M/S compression provides some interesting options, particularly in respect to how you apply compression to different instruments in the mix.

output is radically different, so it's worth distinguishing between the two.

Downwards expansion is predominantly used as a means of tempering low-level noise, and is much the same as what most engineers refer to as a noise gate. Put simply, downward expansion attenuates gain on signals beneath a given threshold. Whereas a compressor 'turns down the loud bits', a downwards expander 'turns down the quiet bits'. In most musical examples, this form of dynamic control isn't particularly useful, but in situations where there are some underlying noise issues – whether it's tape noise, 'open' room mics or some amplifier buzz – a touch of downwards expansion can help keep the master clean and tidy.

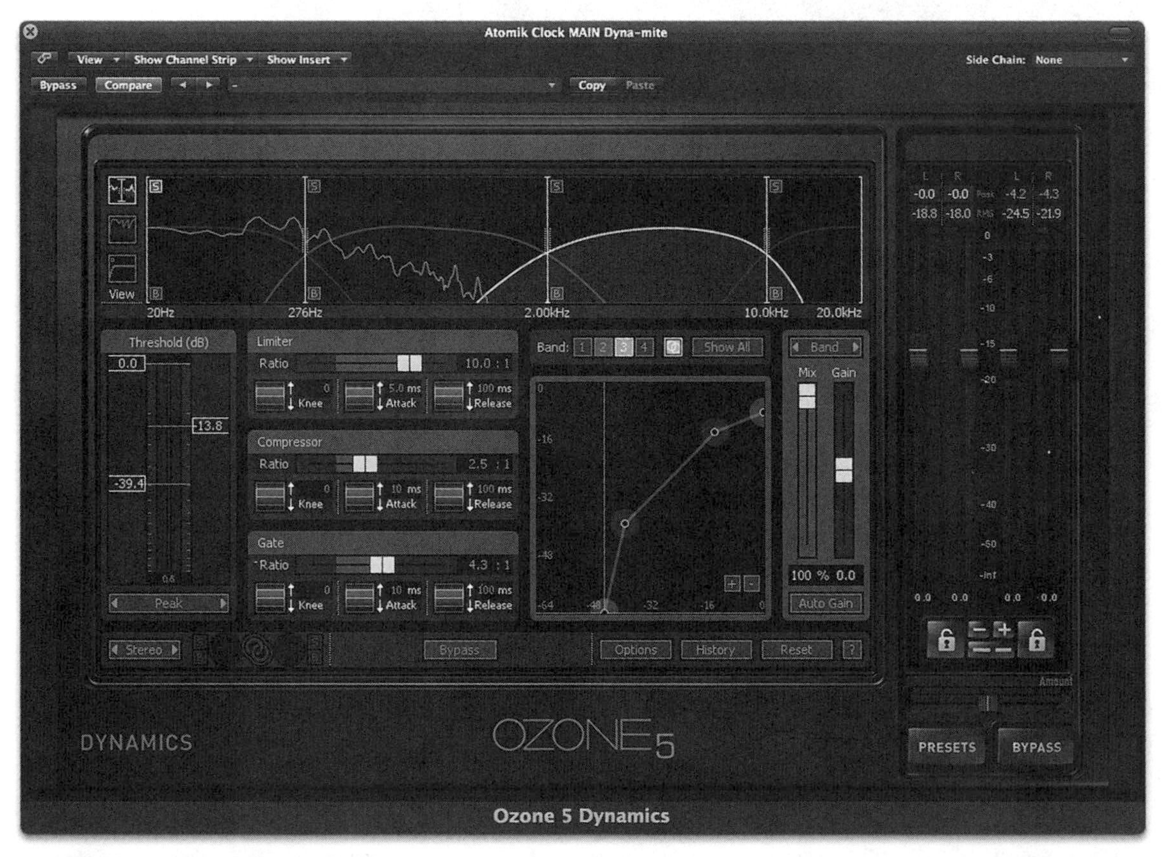

Use a Downwards Expander to control unwanted noise in your master. A ratio of 2:1 should provide enough gain reduction to keep the noise under control.

Setting a downwards expander is much the same as compression, but in reverse. Ideally, the threshold should be positioned just above the noise floor of the track you're mastering, with the ratio defining how hard the attenuation is applied. With a ratio near inf:1, you'll effectively mute the signal when it falls beneath the threshold, which isn't the most musical way of applying the expansion. Using a softer ratio won't attenuate the noise as much, but even 6 dB of gain reduction can make a big difference in the perceived amount of noise in the final master.

Upwards expansion

An altogether different form of expansion comes in the form of an upward expander. With *upwards expansion,* gain is increased as

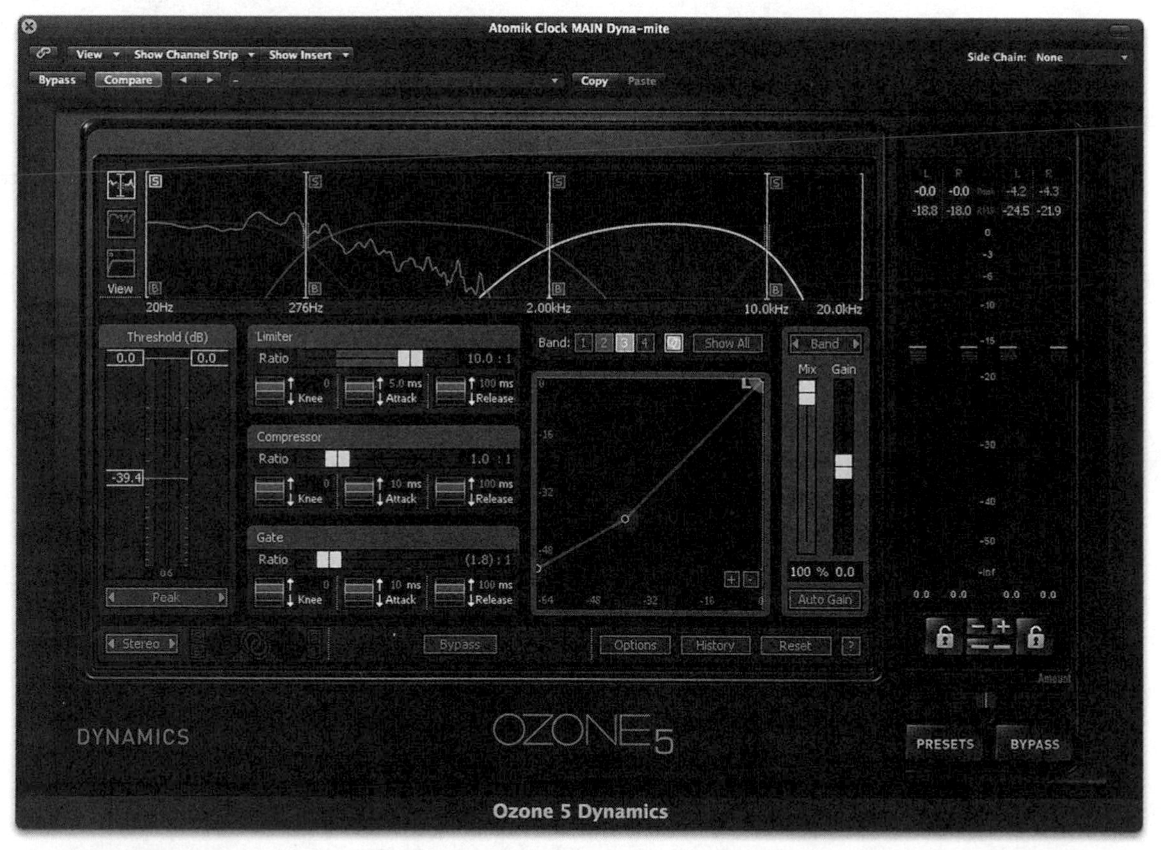

Upwards expansion is an interesting tool for increasing the 'body' of your track, but be careful you don't raise the noise floor in a problematic way.

a signal moves beneath the threshold. Used with care, an upwards expander can help add body to the master, massaging the level of some of the more discrete elements of the mix, although care needs to be taken such that the noise floor (if present) isn't unnecessarily accentuated in this process.

Conceptually, there's some comparison to be made between the sonic output of an upwards expander and that of compressor, in that both can enhance the smaller details and body of the mix. If anything, the concept of compression is much closer to mechanisms in the human ear, so it probably makes most sense to use compression for this type of task rather than upwards expansion. That said, upwards expansion is an interesting dynamic tool, especially if you want to raise the body of the track without the dynamic 'high' sounds being squashed in any way.

CHAPTER 5

Refining Timbre

In this chapter

5.1 An introduction to timbre 92
5.2 Decoding frequency problems 94
Broad colours – the balance of LF, MF and HF 94
Balance of instruments 94
Unwanted resonances 94
Technical problems 95
5.3 Notions of balance – every action having an equal and opposite reaction 95
5.4 Types of equalizer 97
Shelving equalization 97
Parametric equalization 98
Filtering 99
Graphic equalization 99
Phase-linear equalizers 100
Non-symmetrical EQ 100
5.5 A Journey through the audio spectrum 101
10–60 Hertz – the subsonic region 102
60–150 Hz – the 'root notes' of bass 103
200–500 Hz – low mids 104
500 Hz–1 kHz – mids: tone 105
2–6 kHz – Hi mids: bite, definition and the beginning of treble 106
7–12 kHz – treble 107
12–20 kHz – air 108
5.6 Strategies for equalization 109
High-pass filtering 109
Using shelving equalization 111
Understanding the curve of EQ 112
Combined boost and attenuation 114

Controlled mids using parametric EQ 115
Cut narrow, boost wide 117
Fundamental v. second harmonic 118
Removing sibilance 119
5.7 Selective equalization – the left/right and M/S dimension 121
Equalizing the side channel 122
Equalizing the mid channel 124
5.8 Subtle colouration tools 125
Non-linearity 125
The 'Sound' of components 127
Phase shifts 127
Converters 128
5.9 Extreme colour – the multiband compressor 128
5.10 Exciters and Enhancers 130
Aphex Aural Exciter 130
Waves Maxx Bass 131

5.1 An introduction to timbre

Given a contrasting use of instrumentation, differing perspectives and tastes on mixing, as well as varying qualities of recording equipment, it's rare that any two pre-mastered files will ever have exactly the same timbre. Put simply, timbre is one of the big variables of music production – a light 'acoustic' ballad might have an airy timbre with plenty of highs, whereas a hard-hitting piece of dance music will often have a dominant, heavy low-end.

While it's important to retain the individual sonic identity of a track, mastering recognizes that it's also beneficial to have some degree of uniformity to the sound of recorded music, even between stylistically contrasting sources. A deep and pleasantly extended bass, well-rounded mids and nicely defined high-end generally makes for a pleasant and musically effective listening experience, allowing the music to be conveyed in its best possible form. It's also important to understand the restrictions of different playback systems – recognizing, for example, that a deep subsonic bass might not be conveyed across a small pair of speakers, and that not all hi-fi systems extend right up to the airy heights of 20kHz.

Understanding the approach to equalization in mastering, therefore, is often about differentiating between what's possible during mixing and what can be achieved under the surgical precision of a mastering studio. Even with an idea of the 'perfectly balanced' sound spectrum in mind, it's often difficult to ends up with a mix that meets this objective. Try as you might, the bass just ends up being a bit too over dominant, or the treble lacks the brightness and sparkle so often found on a commercial release. Ultimately, the mix is a sum of parts, and those parts might not be adding up in the correct way!

A mix is a sum of its parts, but sometimes the timbral balance doesn't add up in a way that's 100% appropriate.

However, in mastering equalization can be devastatingly effective. Just a 0.5 dB boost at 12 kHz, for example, will affect the entirety of the mix, much the same as if you went through every channel in the mix and applied the same boost. Just a few simple cuts and boosts, therefore, can radically reshape the mix in ways that would be extremely time-consuming (if not impossible) to achieve while mixing. Equally, you can afford to be much more objective about the timbre of a piece of music when you're mastering – spending more time comparing it to other tracks (either on the same CD, or a commercial release) and not having to listen to the same song for hours on end!

As we'll see later on in this chapter, the objective of refining timbre isn't achieved purely through the application of equalization. In truth, there are many ways you can affect the timbre of a piece of music, and although equalization maybe be the obvious and immediate first port of call, there are other ways of achieving similar results, especially where traditional EQ fails to deliver the desired timbral modification.

5.2 Decoding frequency problems

Before diving in with an equalizer, it's worth spending some time trying to identify the specific issues and qualities of the recording you're trying to process. Unlike compression, which tends to deal with broad dynamic issues in a recording, equalization is more directed, often focussing on a specific instrument or a part of the frequency spectrum rather than the mix as a whole. Detailed listening is vital, so that you're clear exactly what you're trying to achieve rather than just 'sweetening' the mix.

To get you started, here's a series of areas that you can consider as a means of exploring any frequency issues that might be present in your recording. Note how they balance both creative and 'colouristic' applications of EQ, as well as corrective tasks that are used to remedy a technical fault in the recording and mixing process. Most if not all equalization moves fall into one of four categories, so this is a good template for making your own analysis.

Broad colours – the balance of LF, MF and HF

Most mixes tend to have a colour, defined by three principle frequency areas: *low*, *mid* and *high* frequencies. A mix that is said to be dull, for example, could be lacking in high-end, or have a disproportionate amount of bass energy behind it. Equally, the mid range could be particularly forced if the mix has been created on a pair of near-field speakers with a poor crossover. Even though there may be other frequency issues at play, the sound of any master can be categorized by the balance of these three key frequency areas.

Balance of instruments

Poor mixing decisions – one instrument being proportionately louder than another – can lead to an imbalance of frequencies and a less enjoyable listening experience. Under the scrutiny of the mastering environment, listen out for problematic instruments that might be making an ineffective contribution to the mix. Strategic EQ – to either boost or cut the frequency area that the instrument sits in – can help restore some sense of balance to the mix.

Unwanted resonances

Unwanted resonances can slip into a recording in a number of ways. Firstly, an instrument itself, or the amplifier it plays through, could have unwanted resonances in respect to how it generates sound. Secondly, the room the instrument is recorded in can have its own resonances, caused by the build up of standing waves. Although a good mix should identify and address these issues, it's often the case that one or more unwanted artefacts slips into the final mix. As the final part of the production process, therefore, mastering is the 'last chance saloon' to address these issues and create a more balanced listening experience.

Technical problems

A range of technical issues in a recording can create various frequency-based issues, including: mains hum, floor rumble, proximity effect, sibilance and so on. As with resonance issues, a good mix should address many of these points, although they can also be remedied by some strategic equalization in mastering.

5.3 Notions of balance – every action having an equal and opposite reaction

The notion of balance is an important concept in equalization, especially when you consider that any action (e.g. a 2 dB boost at 10 kHz) has an almost equal and opposite reaction in another part of the spectrum (in this case, the bass being diminished by virtue of the increased treble). Applying equalization is always a balancing act: understanding how one action might have an effect on a different part of the mix. All too often, inexperienced mastering engineers end up chasing their tails – adding various misplaced boosts across the frequency spectrum when a few strategic cuts might have been more effective.

Understanding this yin and yang of EQ is vital to effective mastering, mainly because the ultimate goal of any mastering engineer is to achieve the desired objective using as few 'moves' as possible. Remember, each cut or boost adds its own phase-shifts, distortion and noise, so the less you have to do, the better the end result will sound. The classic example of this 'minimal moves' strategy is a boost applied to the highs and lows of mix, so as to accentuate the treble and bass respectively. In theory, a similar sound could also be achieved by creating a broad mid-range cut using a parametric equalizer. The immediate advantage here is that you've achieved the same result using just one modification (rather than two), as well as making the whole process potentially more discrete.

Another point to consider with equalization is the overuse of additive EQ and how this affects your objective analysis of a track's timbre and dynamic characteristics. As you'd expect, additive EQ – where the majority of modifications are boosts rather than cuts – increases the overall level of the track. Whenever you bypass the EQ, your point of comparison is both unequalized and possibly 2–3 dB quieter than the equalized version. For this reason, mastering engineers will always try to make a more informed assessment of their modifications by applying a degree of level correction, either on the equalizer itself, or as part of their mastering console.

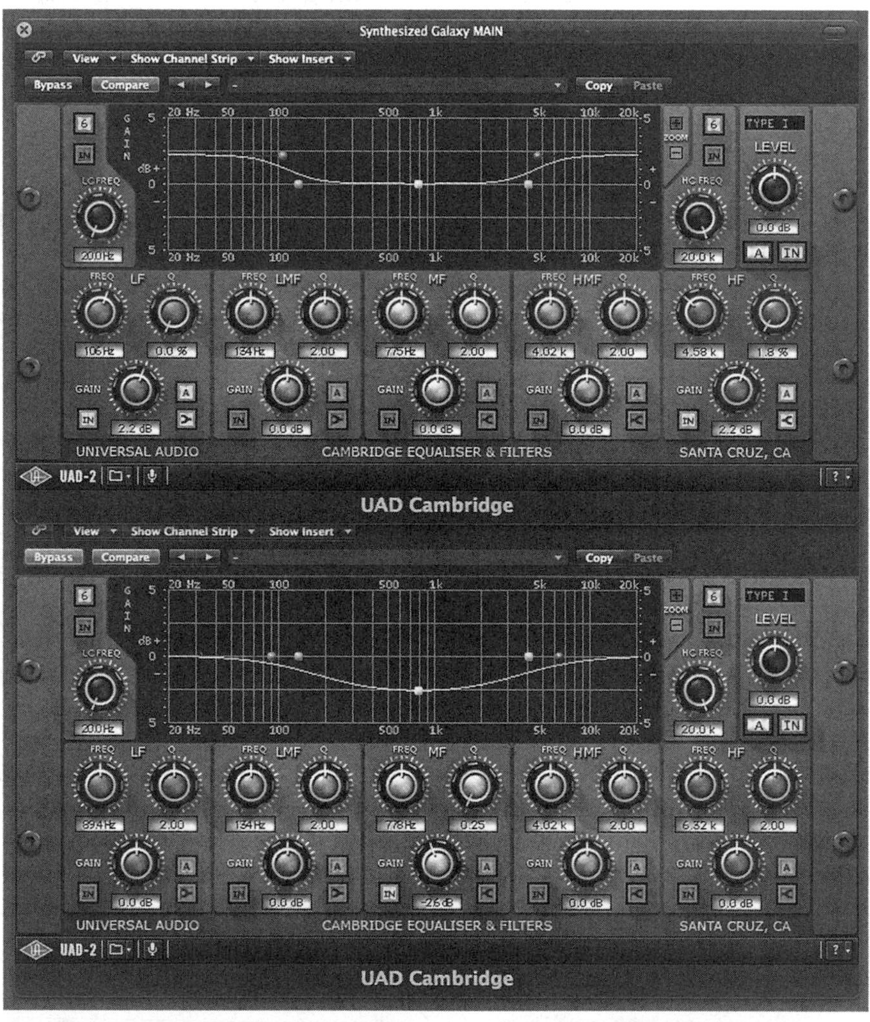

Two contrasting ways of solving the same problem – a wide parametric cut in the middle, and two shelving boosts at either end. But which one is best?

5.4 Types of equalizer

The equalizer is, of course, our principle tool for shaping the timbre of the master. Given its important role, it's no surprise to find a variety of different types of equalizer – from simple shelving EQs, to complex phase-linear designs. The key to effective mastering is often finding the right tool for the job in question, so it's important to make some distinction between the different types of equalization you might have access to.

Shelving equalization

The most common form of equalization is a shelving EQ, which uses two bands (low and high) to modify the two extremities of the audio spectrum. In essence, a shelving equalizer is much the same as the treble and bass controls on a hi-fi amplifier, with just an amount of cut and boost (rated in dB) to modify the input. A high shelving equalizer is generally fixed between 10 and 12 Hz, boosting or cutting all frequencies above this point, whereas a low shelving equalizer will be positioned between 80 and 100 Hz, boosting or cutting frequencies below this frequency. As such, the curve of a shelving EQ is often described as having a 'plateau-like' quality to it – with a gentle rising up or down to a fixed position either above or below the frequency point.

Given that they have few controls and work on the two most easily identifiable components of the audio spectrum (treble and bass), shelving equalizers tend to be the easiest of equalizers to use. However, what a shelving equalizer gains in ease-of-use it lacks in respect to flexibility, effectively limiting you to just tailoring the basic timbral qualities of a sound – whether it's adding a little high-end sparkle or some extra bass-end weight.

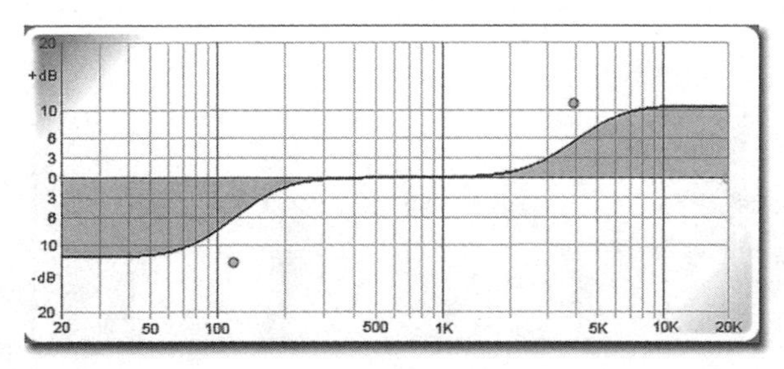

A shelving equalizer is the most common form of equalizer, providing simply tonal control over the highs and lows of a master.

Parametric equalization

An altogether more powerful tool for shaping timbre is the parametric equalizer. To be truly 'parametric' an equalizer needs three parameters: cut and boost (as with a shelving equalizer), a fully sweepable frequency control, and a variable width to the boost/cut known as Q. In short, a parametric equalizer is the 'surgical knife' of mastering – allowing you to position a 'bell-like' cut/boost anywhere in the frequency spectrum (or at least, anywhere between 30 Hz and 16 kHz). By varying the Q an engineer can adjust the width of the boost – either allowing them to tune out a specific rogue harmonic, or apply a broad cut/boost across a range of frequencies.

To make the equalizer easier to negotiate, a fully parametric design tends to offer two or three bands of equalization, each covering a proportion of the audio spectrum. This is usually denoted as *low-mid frequencies* (or LMF for short), and *high-mid frequencies* (HMF). On a more advanced equalizer it's common for shelving controls to take on certain 'parametric' qualities – whether it's the ability to adjust the frequency of the shelving controls, or to vary the Q. In the case of variable Q, the adjustment adapts the relative shape of the 'plateau-like' curve, moving it between a gentle ramp-up/down at the frequency point, to an altogether steeper curve.

Of course, not all equalizers meet the criteria of fully parametric operation. Although a sweepable frequency control might be provided, as well as an amount of cut and boost, it might be that the Q is fixed at a predetermined bandwidth, or switchable between a tight Q setting and a wider Q. Older equalizers might also omit the ability to sweep the frequency control, and instead provide a number of predetermined frequency settings. Although the omission of fully variable Q and frequency controls might seem like a compromise, these limitations also lend the equalizer a distinct character and charm.

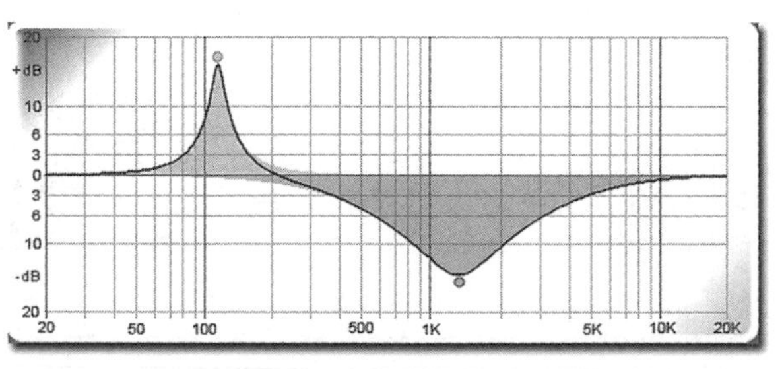

Parametric equalizers provide a more detailed set of control for modifying timbre. They are particularly useful working on the mid range of your master.

Filtering

Positioned at either end of a fully fledged equalizer you'll find two filter controls. Unlike shelving or parametric EQs – which can apply either a cut or a boost – a filter attenuates all frequencies either above or below a given cutoff point. A low-pass filter, for example, lets frequencies below a given cutoff point pass unmodified, while frequencies above the cutoff point are cut. A high-pass filter works in reverse, filtering out low-frequency signals, while leaving the treble in its unmodified form. In effect, a filter allows you to top-and-tail the harmonic content of sound using broad brushstrokes This provides a quick-and-easy way to remove excessive rumble, for example, or to make a sound distinctly darker in timbre.

Alongside a frequency control, a filter will often provide a strength control, and occasionally a variation on Q. The strength of the filter is measured in decibels per octave, with a 6 dB curve being relatively shallow, and 12 dB or 18 dB curves producing a sharper cut. Adding Q produces an emphasis around the cutoff point, ahead of the attenuation. On a synthesizer (which uses a filter as a component in its sound-shaping signal path), Q is also referred to as resonance.

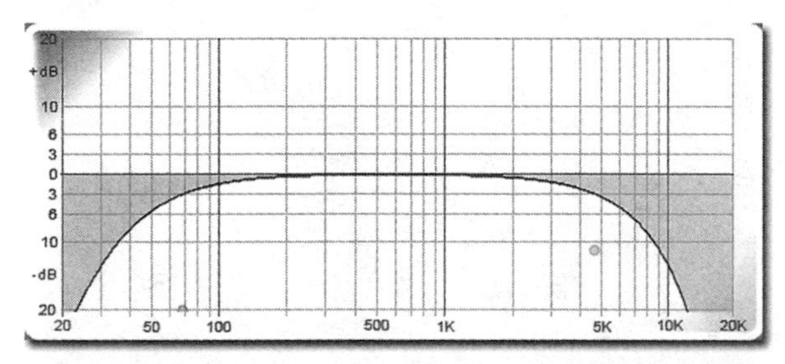

Filters are extreme tools, providing sharp attenuation above or below a given cutoff point.

Graphic equalization

An interesting variation of parametric equalization is a so-called graphic equalizer, which provides a discrete control for up to 31 bands of equalization. Despite being a useful tool with respect to being able to visualize the overall curve of equalization, a graphic EQ is often a clumsy and inefficient tool for mastering. Firstly, all the bands have a fixed Q setting – which is good in that it allows each band to have a unique control over its portion of the frequency spectrum, but also makes it impossible to produce broader boosts or cuts without the individual peaks standing out. Secondly, although 31 bands might seem adequate, it still remains a relatively 'coarse' division of the audio spectrum, with each control effectively covering 1/3 of an octave (which is why a 31-band device is referred to as a 1/3-octave equalizer).

Phase-linear equalizers

A more recent development is the so-called *phase-linear* equalizer. In theory, a traditional analogue equalizer will produce small phase shifts wherever a cut and boost is placed in the frequency spectrum. However, a digital phase-linear equalizer can negate the phase-shifted artefacts, although this can be at the expense of additional processing latency.

Technically, a phase-linear EQ preserves the signal integrity much more than a traditional equalizer – and certainly comes in useful when equalized and unequalized signals are merged – but it's also interesting to note that these phase shifts positively contribute to the EQ flattering the sound you're trying to process. In short, the phase-shifting is almost as integral to the 'sound' of the EQ as the amount of cut and boost itself, and shouldn't be thoughtlessly discarded in the search for technical perfection.

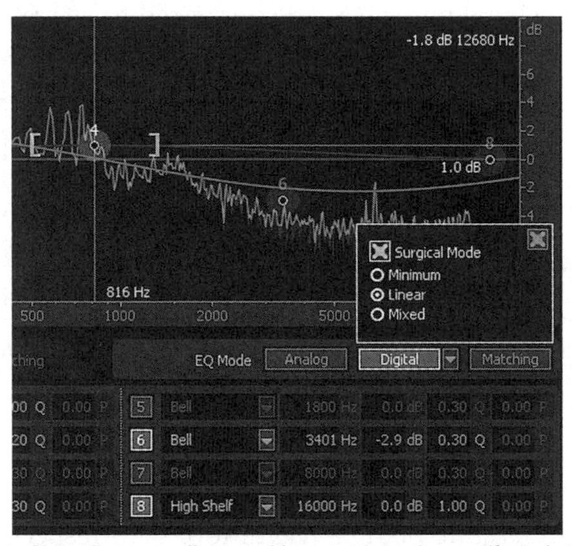

As the name suggests, Phase-linear equalizers avoid unnecessary phase artifacts, but this might be at the expense of the 'sound' of a good EQ!

Non-symmetrical EQ

As we saw previously, one of the key parameters in a parametric EQ is Q – the width of the cut or boost. In a theoretically perfect equalizer there is a completely linked and symmetrical response to Q, so that whether the equalizer is cutting or boosting (or varying amounts of cut and boost are being applied) the width remains constant. This seems to make a lot of sense at first, but is it the most musical response, or the one that is empathetic to your objectives as an engineer? For example, if you're adding an increasing amount of boost, doesn't that suggest that you're interested in a more specific, rather than broad, set of frequencies? Ultimately, the best form of equalizer is one that adapts itself – or more specifically, the width of its boost – to intelligently match what you're doing.

For this reason, you'll find that a large number of vintage EQs have some interdependency between the width of cut/boost and the amount of gain applied, and, in some cases, a 'nonsymmetrical' response. For example, Neve EQs are known to have an increasingly sharper curve as gain in increased – focussing the EQ, in effect. A decreasing width with gain also makes more sense in respect to a more consistent amplitude with differing gains. A wide curve on a low gain setting, for example, would have a negligible effect on overall amplitude, but increase the gain and the same broad curve could bring the overall signal level up by a considerable amount. In effect, this Neve-type approach makes it far easier to experiment with different gains without having to adjust to large leaps in overall signal level. The only downside might be the inability to use small, precise EQ settings (for discrete removal of a few surrounding frequencies), which is why tight parametric EQs – such as the SSL 4000 series – are still popular equalizers to have to hand.

Nonsymmetrical EQs are another variation, differentiating between the desire to add broad colouristic boosts to an EQ setting, and the need to notch-out problematic frequencies such as instrument resonances, mains hum and so on. A nonsymmetrical EQ – such as the Helios Type 69 – will have a bias towards wide curves in its boost phase, while corresponding cuts are sharper and more pronounced. Again, the idea is that a nonsymmetrical operation should make the EQ's response more musical, but it does mean that boosts and cuts can't be easily reversed without some corresponding adjustment in your Q setting.

5.5 A Journey through the audio spectrum

Although every piece of music presents its own unique spectral profile, it is worth noting the unique attributes and qualities of the different parts of the audio spectrum. Indeed, even the most casual of listeners will be aware of the two main spectral areas of music – treble and bass – but there's a far greater palette of frequency areas to understand, each with their own positives and negatives that can either aid or hinder the effectiveness of a master. A vital role of the mastering engineer, therefore, is to fully understand and appreciate what these areas can deliver.

What follows is a step-by-step guide to the audio spectrum, starting off with the lowest frequencies and then working up to the extremes of human perception. Within reason, we've kept the areas relatively broad, so it's worth remembering that there are even finer divisions within each of the sections described.

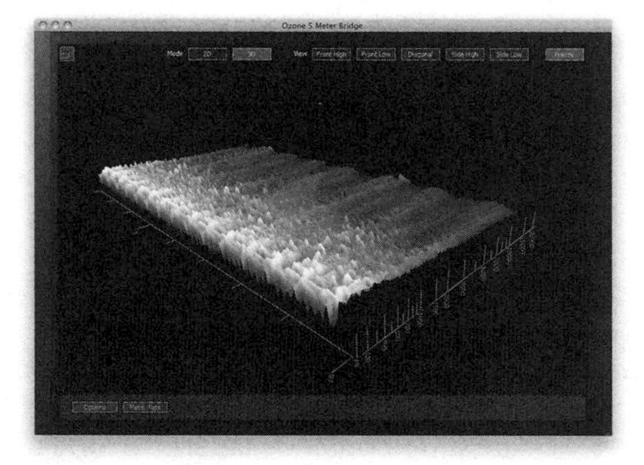

The complete audio spectrum contains a diverse range of frequencies, each with a different role to play.

10–60 Hertz – the subsonic region

Key sounds: deep 808 kick drums, dub bass, subsonic rumbles.

The subsonic depth of any mix is contained in the 10–60 hz region – an area that is more of a physical sensation rather than an audible response to sound. In short, our ears have difficulties discerning subtle differences between frequencies down in the subsonic region, but given the amount of sound energy used to make the low-end audible we certainly appreciate its presence! The subsonic region is all about depth and power – but hopefully applied in a controlled and appropriate way.

From a mastering engineer's perspective the first big issue is that of control. The amount of sound energy used to generate such deep low-frequency sounds can often eat up your available headroom – the space that could easily be taken up by other, more audible parts of the mix. It's also highly probable that the end user's hi-fi system simply won't be able to output such low-frequency signals (this is certainly the case with most cheap computer-based speakers), so any energy expended is this area somewhat wasted. Finally, of course, there's also the issues of uncontrolled low-end slipping into the recording – whether it's proximity effect, floor rumble or any other undesirable low-frequency sound that might not have been noticed initially.

Given this need for control at the extreme low-end of the master, an almost compulsory EQ setting is a high-pass filter strategically positioned to keep the subsonic region under control. Of course, a decision on the position (usually around 20–40 Hz) should be formed from an understanding of your source material and the potential bandwidth of your desired format. To best judge your input source requires some full-range monitoring, so be wary of making hasty decisions if your monitoring setup doesn't extend as far as your equalizer. As regards playback, you may want to consider the relative merits given your end destination. As one example, a soundtrack on TV has different sonic objective to that of a film soundtrack played in the cinema. Differentiating between these two applications, therefore, might best ensure your music is presented as effectively as possible.

60–150 Hz – the 'root notes' of bass.

Key sounds: kick drum, bass guitar, synth bass.

Moving above the subsonic region, the next area we encounter is largely formed by the fundamental frequencies of our bass instrumentation – the kick drum, bass guitars and bass synths. Unlike the subsonic region, this area of the frequency spectrum forms a vital part of a song's musicality, effectively forming the harmonic foundation that the rest of the track sits on.

From a mastering perspective, it's vital that the 60–150 Hz region is solid, well balanced and proportionate to the rest of the mix. All too often a mix will be delivered with an overcooked bass-end (this is often a result of bass instruments being driven too loud in the mix, or an over enthusiastic application of EQ), or a general sloppiness where the levels of bass are inconsistent. Where the bass levels are too high, the track will exhibit a 'heavy' overall sound, as well as potentially sounding muddy.

Another point worth noting with bass is our inability to decode too much information in that area. Put simply, we have difficulty discerning pitch at the bottom-end of the frequency spectrum, which is why it is easier to spot intonation problems on a guitar rather than a bass. As a result, good arrangers have used a 'less is more' approach to instrumentation in the bass spectrum – using just one or two key instruments (usually kick drum and bass guitar) rather than any complicated layering effects. When it comes to mastering, therefore, we need to make sure our bass spectrum is clean, ordered and well focussed, which might mean that we have to call upon a number of signal processing tools and not just equalization.

As well as the obvious bass instrumentation, it's also interesting to note how other sounds can 'leak' into this frequency area, which can sometimes be a good or a bad thing. Weighty snare drums, booming electric guitars and 'sloppy' synthesizers can start to creep into this area. If positive, this will add power to the recording, but can also make it muddy and indistinct. Of course, applying mastering EQ to these instruments will also affect the bass, so it's well worth addressing some of these issues in the mix if at all possible.

Your bass-end needs to be tight and controlled. Keep an eye on an overcrowded bass, and ensure your track doesn't become too heavy through excessive amounts of bass energy.

200–500 Hz – low mids

Key sounds: piano, 'low' guitar, low strings, french horn, toms, snare (power).

Low mids are an interesting part of a mix, being an important part of the warmth, weight and body of a track. However, applied in an uncontrolled way, a track can soon become muddy and slightly stodgy to the ear.

From an arrangement perspective, it's pertinent to note the amount of instrumentation that sits in this part of the frequency spectrum. Guitars, pianos, french horns, and low strings can all start to clog-up the 200–500 Hz part the frequency spectrum, and if lots of close mic-ing has been used to record these instruments, there's also likely to be a degree of proximity effect that will add extra weight to the low mids. Arguably, there's a strong case for some well-honed arrangement skills here as well as a touch of strategic high-pass filtering on individual channels to keep the low mids under control. When it comes to mastering, however, we need to take an objective view on this part of the frequency spectrum, and adjust it as necessary.

Low mids are important for warmth, but can also make a track muddy. Look out for instruments that extend their frequency range down in this region – including Pianos, the low-end of Guitars and French Horns.

500 Hz–1 kHz – mids: tone

Key sounds: vocals, guitar, strings, snare.

From around 500 Hz to 1 kHz a track has to carry a large amount of musical information, which is no surprise given that our hearing is most sensitive in this part of the frequency spectrum, and that the two most important octaves in music – the notes between C3 and C5 – sit principally in this area.

Although there's a lot of pitch information in the 500 Hz to 1 kHz area, it's not an area that excites many listeners. When the mids become too forced, the sound becomes slightly nasal, and as a result you loose some of the vibrancy and energy of an extended bass and treble. That said, the mids are still an essential part of how an instrument articulates and communicates its message, and it may well be that a strategic boost in this part of the spectrum offers the small but important lift to an instrument that needs to be heard.

A lot of key musical information – including the vocals – is conveyed between 500Hz to 1Khz, but this isn't a frequency area that's greatly to do with the colour or timbre of instruments.

2–6 kHz – Hi mids: bite, definition and the beginning of treble

Key sounds: vocals, guitar, snare.

As we rise above 2 kHz, some interesting and useful sounds start to slip into the equation, largely because we're starting to deal with harmonic overtones rather than discrete 'notes' (for reference, the fundamental of the highest note on a piano is around 4 kHz). Given that we're mainly dealing with overtones, we now start to adjust the tone and colour of our recording, rather than directly manipulating the 'instrumental' parts. In particular, the 2–6 kHz spectrum is where we perceive a large part of the bite and definition of a recording – not treble as such, but the beginning of brightness and certainly an important part of the extended timbre of many principle instruments.

Given the bite of the 2–6 kHz spectrum, it has particular relevance for key instruments such as vocals, guitars and the snare. Adding a boost in this part of the spectrum really pushes these instruments through in the mix, adding drive and intensity. Too much boost, though, and the 2–6 kHz frequency range can sound too aggressive and pushed, often at the expense of the natural extension and brightness that is to follow above 7 kHz. Any distorted guitars are often rich in this part of the frequency spectrum, due to the added saturation and the natural roll-off that occurs above 10 kHz on a cabinet speaker. Again, it's a question of balance – either adding boost to help the articulation, or taming an excessive use in this area.

Another interesting issue with the 2–6 kHz spectrum is the potential 'hole' effect of a poorly aligned near-field crossover. Where a near-field speaker has a 'less than flat' mid response, you can often end up with mixes that are slightly over-pronounced in this 2.5 kHz area. This is a perfect example of why a three-way set of active monitors is often the preferred choice for mastering, as this will always highlight any problems with the mix in this crucial part of the frequency spectrum.

7–12 kHz – treble

Key sounds: hi hats, cymbals, acoustic guitars, transient details.

Other than bass, treble should be another immediately recognizable part of the frequency spectrum, forming a defining ingredient in the timbre of a track. The sizzle of a cymbal, the sheen of a new set of steel strings on a guitar, or the transient detail on a snare drum. In short, treble provides much of the detail in a track, and is an important part of the listening experience. By slightly forcing the treble we can also create a unique form of auditory 'saturation' much like the forced colours on a Bollywood film. In effect, the increased treble enhances the detail of the recording, increasing its vibrancy and energy.

However, pushing treble too hard has some distinct downsides, often making a recording seem unnecessarily harsh on the ears, as well as highlighting any problematic noise in the source recordings. It's also important to remember that an equalizer can only boost what's present in the original source material. For example, if the recording sounds dull and has no 'recorded' high-frequency detail, then no amount of equalization will generate new harmonic information to 'fill the gaps'. Likewise, if the source is already sounding aggressive, applying HF equalization will simply make matters worse.

Within the treble frequency spectrum there are also a great deal of different colours to play with. While most engineers plump for 10 kHz as the default frequency, you can also extract some interesting sonic variations either side of this line. 7 kHz, for example, is a useful place to start a lift if your source is lacking in HF detail. 12 kHz, on the other hand, has a lighter sound to it, by virtue of it being slightly higher up the audio spectrum. If your source has plenty of HF extension, a 12 kHz boost has a cleaner sound, enhancing the detail with a little less 'edge' than at 7 kHz.

Starting off from around 7kHz, we find the real 'detail' of a recording. Transients are often dominant here, particularly on percussive instruments like drums.

12–20 kHz – air

Key sounds: strings, cymbals, acoustic guitar.

As with the subsonic area, the extremes of treble – also known as *air* – are closer to being a physical sensation rather than something you distinctly hear. Even though your hearing becomes significantly attenuated above 10 kHz there's still plenty of frequency material dancing about, all of which seems to create a subtle 'fairy dust' on top of the mix. Controlling the articulation of the extreme highs in a mix is an important way of defining its tone and colour. On well-recorded acoustic music, an air lift can often help enhance its 'life-like' qualities, emphasizing the important details that sit at the top of the audio spectrum. On less naturalistic sources, an air boost is often less helpful, or in some cases of negligible impact given the lack of any true 'highs' in the source material.

Although the stated bandwidth of human hearing stretches from 20 Hz to 20 kHz, there's some interesting debate as to the merits and contributions of frequencies above 20 kHz. This particularly explains the popularity of running at higher samples-rates – such as 96 kHz – and why many audiophile plug-ins use upsampling. For mastering, although you might deliver at 16 bit, 44.1 kHz resolution, there's a strong argument for running at higher resolutions for the purposes of signal processing, especially if you're using analogue equipment, which has no upper restrictions placed in its bandwidth. Either way, there's clearly some magic happening above 12 kHz, so we need to ensure that we preserve and enhance this detail accordingly.

Above 12kHz we largely start to deal with the 'air' of a recording, which is often in relation to the sheen of instruments like strings.

5.6 Strategies for equalization

High-pass filtering

For some, a strategically positioned high-pass filter as the first component in your signal processing chain is a near-essential part of mastering. In truth though, there's a fair degree of debate on this issue, and it's well worth reading our description of the 1–60 Hz region as part of our 'journey through the audio spectrum' to see the musical and artistic merits of it, as well as the key technical issues (including the use of sound energy and the bandwidth of playback systems) that it poses.

The biggest driver regarding what you apply in respect to high-pass filtering will be the actions already carried out in the preceding parts of the recording process. In short, if the recording and mix engineers have done little to control the subsonic region – either by forgetting to place a bass roll-off on offending microphones, or by neglecting to deal with the bass-end as part of the mixing process – then it's important that the process of mastering makes a balanced and rational decision as to the presence of sub bass. Even with some degree of control, there might still be an argument for an extra degree of restraint as this can help the fundamentals of bass (found above the 10–60 Hz region) articulate themselves in a more effective way.

Most mastering equalizers have a fixed strength setting for the high-pass filter, which generally veers towards a steep attenuation rather than a gentle slope. If you're using a generic software equalizer, try to ensure that the slope is steep rather than shallow (this will generally mean opting for a higher dB per octave setting) so that only the lowest frequencies are attenuated, and all frequencies above the cutoff-point pass largely unaffected.

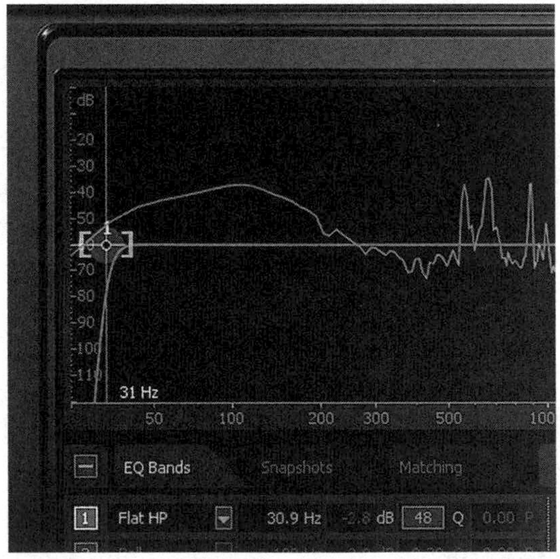

A high-pass filter will keep the subsonic frequencies of your track in check, but ensure it doesn't kill the power of your master.

Using shelving equalization

Being relatively broad tools, shelving equalizers are a great way of changing the colour and tone of your master. As with the application of high-pass filtering, it's well worth familiarizing yourself with the different musical colours that occur over the bass and treble component of the frequency spectrum, as detailed in our journey through the audio spectrum. Arguably the biggest mistake with mastering is to simply opt for the standard shelving settings (100 Hz and 10 kHz), which tend to deliver predictable and usable results, but might not be 100 per cent appropriate for the particular timbral fingerprint your track is presenting. Often the 'right' setting is found through a slight deviation from these standards – either lifting a high shelving boost up to 12 kHz, or pulling the bass slightly lower to 80 Hz.

Unless you're faced with a particularly 'skewed' frequency balance, most shelving boosts or cuts will tend to sit in the 0.5 dB to 2 dB range. This might seem slight at first, but considering that all frequencies above or below the cutoff point are modified by this amount, your actions can have quite a big impact on the sound of the master.

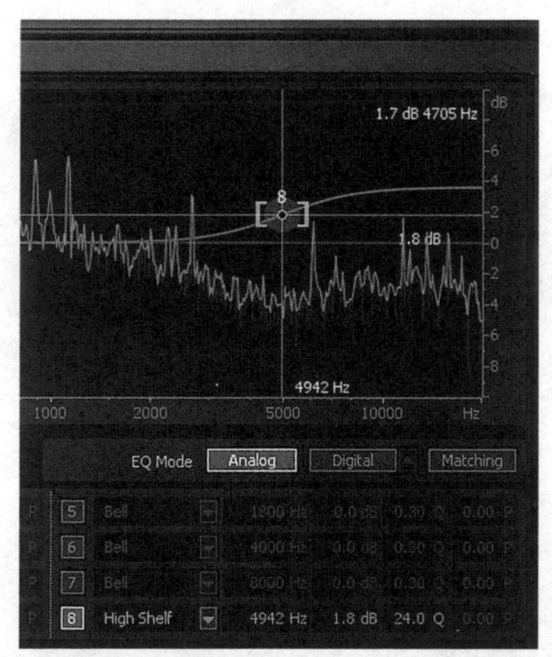

Basic shelving settings allow us to influence the overarching colour of the master, as well as adding a small amount of 'hype' to the sound.

Over and above the 2dB limit, you're starting to take the modification into deliberately 'hyped' settings, which may or may not be to the detriment of the music. As a broad guide, acoustic sources tend to be less tolerant of such drastic modifications, whereas electric or electronic sources can often handle greater amounts of 'colour' enhancement. Be wary though, because your ear can soon adjust itself to these extreme settings and start to hear the curve as being 'normal'.

Understanding the curve of EQ

One of the most important parts of shelving equalization to understand is its curve, as this will have a big effect on the precise nature

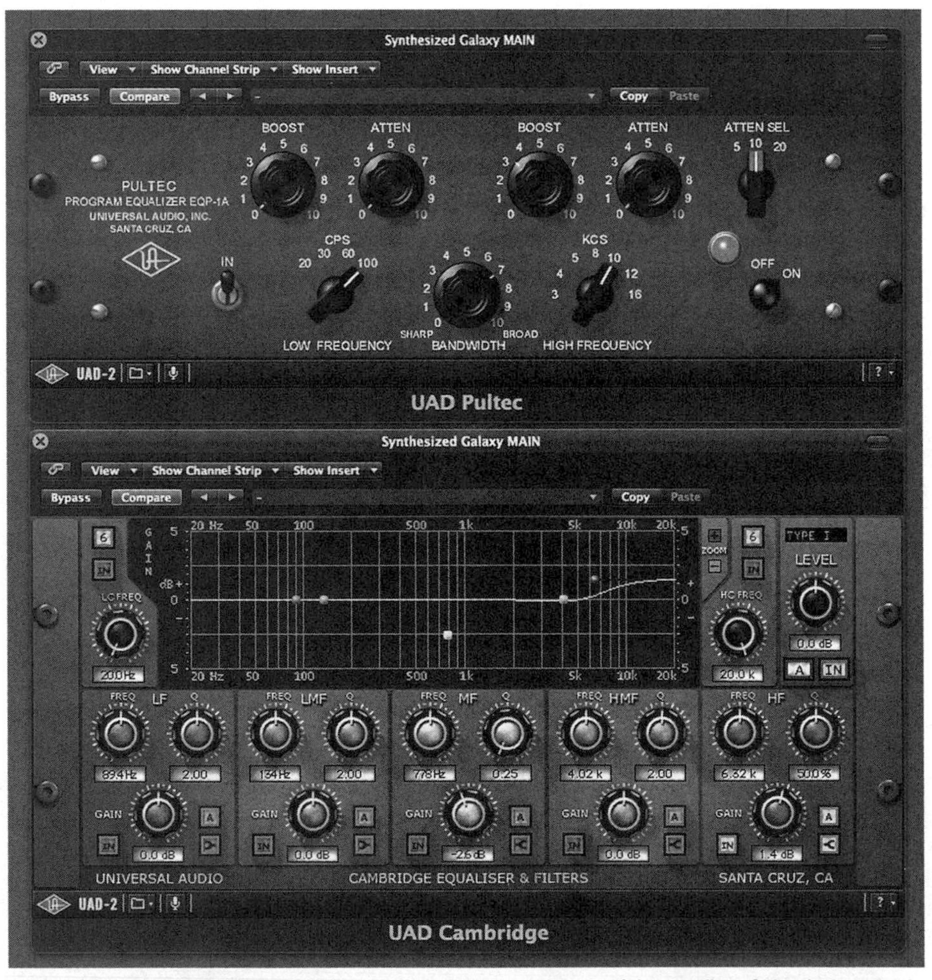

The precise colour of a shelving boost is largely defined by the curve of the equalizer. Older designs tend to exhibit wide, musical curves, whereas more contemporary designs feature a precise 'plateau' effect.

and quality of the cut and boost you apply. Put simply, there are some big differences in the curves of different shelving equalizers, with some equalizers exhibiting a sharp plateau-like curve, while others have a more rounded slope. These different shapes trade-off surgical accuracy against character and musicality, allowing you to make an informed decision between the two different approaches.

The precise nature of the curve is defined either by the design of the EQ, or by adjustments to the Q setting in situations where the equalizer has a fuller set of controls. Older designs of equalizer, such as the Pultec EQP-1A Equalizer, often featured curves with wide slopes, which tended to sound relatively smooth on the ears but were also much less precise about the frequencies that were brought up as part of the boost. An extreme of this 'wide curved' shelving is the so-called Baxandall EQ, originally developed as a tone control for hi-fi systems

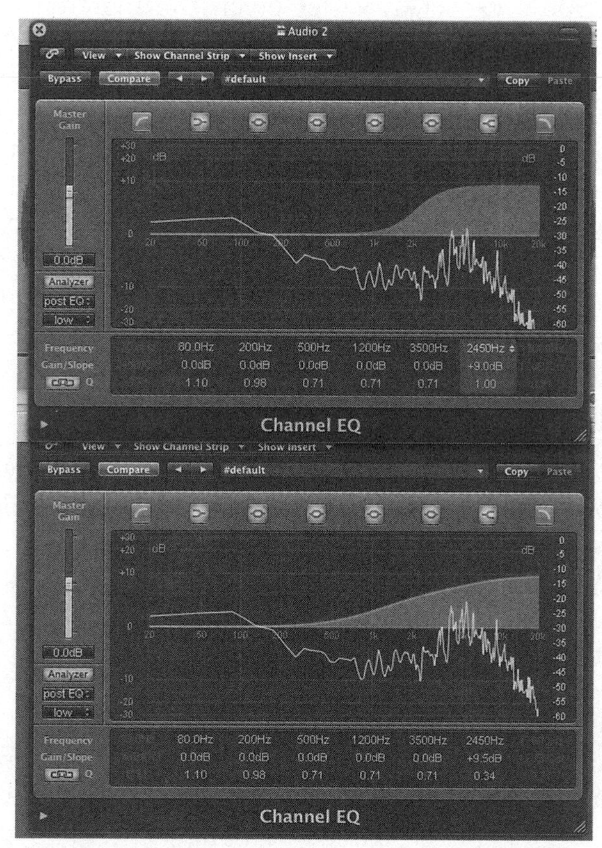

By varying the Q of a modern parametric equalizer you can often adjust the shelving's curve.

back in the 1960s. The Baxandall curve appears to omit the shelf altogether, and simply applies an increasing amount of boost (or attenuation, for that matter) as it goes up or down the frequency spectrum.

For a more refined form of control, shelving equalizers were developed with a distinct plateau, so that only frequencies above or below the cutoff point were modified. Having a more defined curve allows you to be more selective with your equalizer – so that only the 12 kHz 'air' is lifted, for example, rather than pulling up a collection of treble frequencies starting from around 7 kHz. Modern shelving equalizers that feature a Q control offer the potential of varying the slope, arguably moving from a wide Baxandall-like curve to something rather more precise and accurate. You can also use wide parametric boosts positioned high up in the frequency spectrum (usually around 16 kHz) to emulate the gentle rising effect of a Baxandall boost.

Of course, the question we then need to ask is what curve sounds the best? As you'd expect, it varies given the material you're working with and the effect you want. Try contrasting a few equalizers, applying the same shelving boost to see what sounds best. As well as listening to the obvious principle boost, though, listen to what happens to frequencies just above or below the cutoff point – are some adjacent instruments being modified for example, and is it musically appropriate? Gauging both the positive and negative attributes of the EQ will allow you to make the most informed decision, with the best result possibly not being the first equalizer you immediately plump for!

Combined boost and attenuation

This combined approach of a low-end cut and boost, which simultaneously controls and enhances the bass end, is well illustrated on two unique signal processors – the Pultec EQP-1A Equalizer and Little Lab's Voice of God Bass Resonance Tool. The Pultec's lowest frequency band (which can switch between 20, 30, 60 and 100 Hz) has individual controls for both boost and attenuation. When both are combined you end up with a curve with both bass attenuation (slightly above the stated frequency) and bass boost (below the stated frequency). Little Lab's Voice of God Bass Resonance Tool, on the other hand, uses a form of resonant high-pass filtering so that there's a balanced application between subsonics being cut and bass being accentuated. With a parametric equalizer, you can achieve a similar effect by pushing the Q harder so that you see a small dip ahead of the cutoff point.

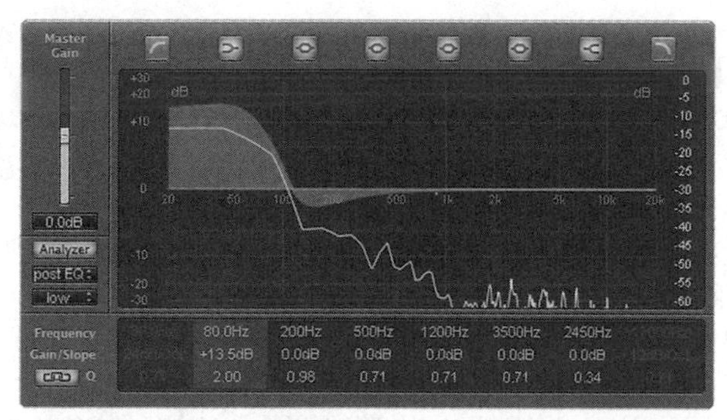

Combing boost and attenuation around the cutoff point can add an interesting colour to the EQ. In this example, the Q has been pushed harder to replicate the sound of a Pultec Equalizer.

Whatever tool you use to apply it, the sonic results of a combined boost and attenuation tends to be a 'best of both worlds' sound – so that, for example, we hear an increased amount of bass presence, but also a slightly reduction in the low mids to counterbalance this enhancement. Arguably, the strategic dips also make the modifications more distinct, almost appearing to 'clear the way' ahead of the boost taking place. Of course, it's not an effect that suits every form of music, but it's an interesting example of a form of yin and yang EQ in action, and important for understanding how a balanced combination of moves can add up to create a greater overall effect.

Controlled mids using parametric EQ

Having dealt with the broad colours of our master, parametric EQ offers the ideal opportunity to influence the smaller (but equally important) details of the music. Parametric EQ offers an unprecedented amount of control over the music, arguably allowing us to fine tune the master on an instrument-by-instrument basis. Even so, it still gives you the opportunity to work with broader tonal properties, particularly how the mid range is articulated and presented in the final mix. Indeed, even small changes in the mid-range can change the perceived bite, clarity, warmth or depth of the final master.

Given the three principle controls available on all parametric equalizers – cut/boost, frequency and Q – we need to develop a slightly different way of using them. First and foremost, we need to locate the given frequency that we want to work with, which can either be achieved through an innate understanding of the frequency spectrum, or by using the parametric EQ to 'tune in' to the given frequency.

Finding the right frequency is a good skill to develop with a parametric equalizer, and will certainly improve your understanding of the audio spectrum. You need to start by adding a relatively harsh amount of boost using a narrow Q setting – just enough so that a small range of frequencies are brought into distinct 'focus', but not so harsh that your ears hurt! Now move the frequency control through its given part of the frequency spectrum, trying to find the precise frequency to want to work with. This frequency could be a specific instrument in the mix, a colour within the mid-range, or an unwanted resonance. Either way, you should be able to find the frequency you want using this sharp boost.

Once you've found the desired frequency, you can then start to adjust the amount of timbral modification using the Q and cut/boost controls. Start by resetting the cut/boost to 0 dB (so that your

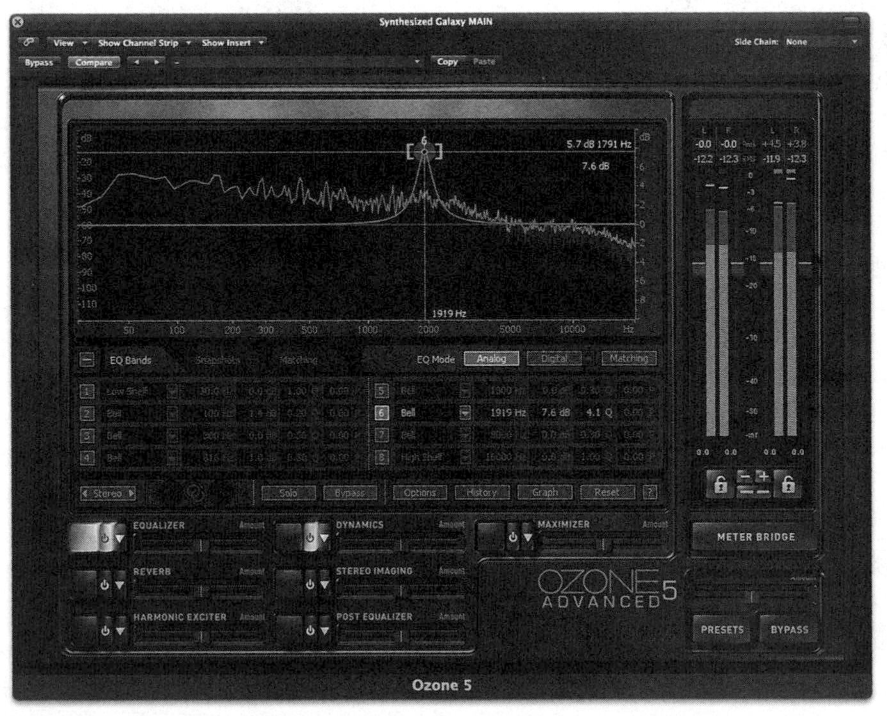

Use an exaggerated, narrow boost to find the precise frequency you need to work with.

ears can reset to a normalized state) and widen the Q setting so that it's working with an average-sized curve. Once you've found the reset point, start applying the cut or boost as required, either widening the Q to increase the scope of the boost/cut, or narrowing it to focus the EQ on a specific instrument or frequency.

Cut narrow, boost wide

If in doubt, a good overarching strategy with parametric EQ is to cut using a relatively narrow Q, but boost using a wide Q setting. The methodology employed here makes some degree of sense given your objective when using a parametric EQ. When you're cutting with a parametric equalizer it's often because you need to attenuate a specific problem – maybe a harsh bass resonance, or an instrument that's too loud in the mix. In this situation, keeping the Q narrow will keep the modification relatively discrete and unobtrusive, as well as allowing you to increase the amount of cut without affecting the adjacent instrumentation.

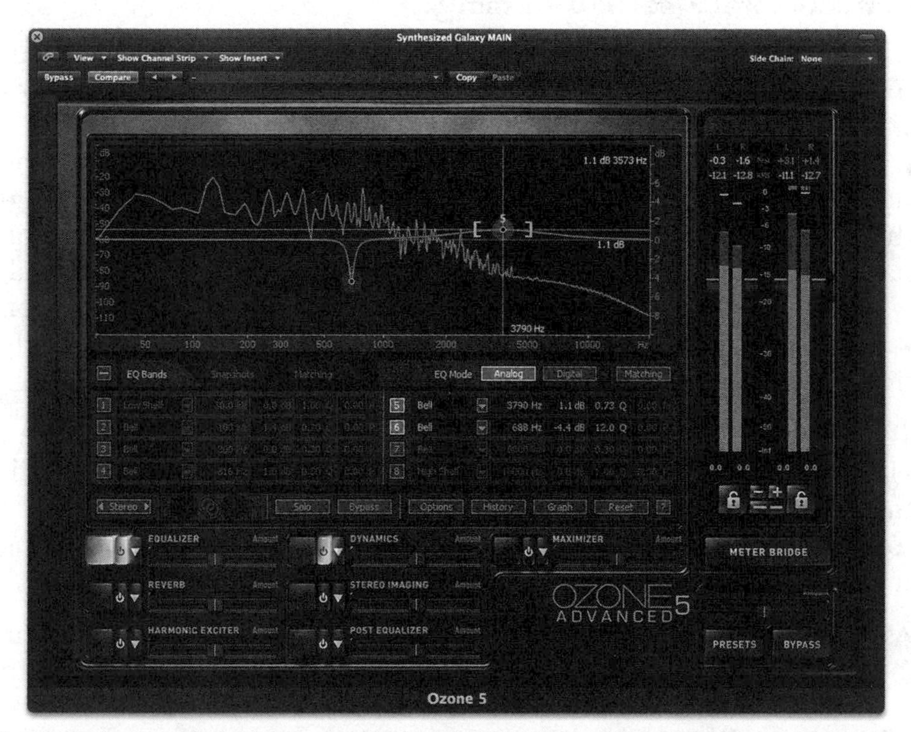

For the most transparent application of parametric EQ, look towards applying wide boosts and narrow cuts.

Parametric boosts, on the other hand, are often to do with the colour of the recording – perhaps adding a touch of presence around 5.6 kHz, or a little more warmth around 300 Hz. In this case, you're much better using a relatively gentle amount of boost on a wide Q setting so as to keep the effect discrete. Indeed, when it comes to boosts, the use of a narrow Q setting tend to be particularly noticeable, which is why it tends to be so useful for 'tuning' the parametric's frequency control, but slightly less useful if you want your equalization fingerprint to be light and transparent.

When it comes to the colour of a recording, a parametric equalizer can apply some transformative processing on a master. Rather than repeat ourselves, it's worth referring back to Section 5.5 and our 'journey through the audio spectrum' to see the attributes that can be found in the various parts of the mid range. With experience, you'll soon find how different balances within the mid range can significantly affect the articulation and presentation of a piece of music, and how a good dose of parametric EQ can address many of these issues.

Fundamental v. second harmonic

In situations where you want to address the balance of an instrument in the mix, or change how an instrument is presented, it's worth considering its harmonic structure. All sounds are constructed from a fundamental frequency (what we perceive as the pitch of the note) and an ascending series of pitched overtones – or harmonics – that rise above the fundamental. Sounds that are said to be 'musical' have harmonics that are mathematically related to the fundamental, while sounds that have a degree of inharmonicity (such a drums, or, in a subtle way, bell sounds) have harmonics that deviate slightly from an exact multiplication of the fundamental.

To see the structure of a given sound it's well worth looking at an FFT (Fast Fourier Transform) plot, which shows the respective amplitude and distribution of harmonics. The fundamental is always the loudest frequency on the display, with the rising harmonics falling in amplitude as you rise up the harmonic series.

As we hinted in our 'journey through the audio spectrum' (Section 5.5), it's interesting to see how we apply equalization with respect to an instrument's harmonic structure, as well as in the context of the massed harmonic structure of a full ensemble. When we want

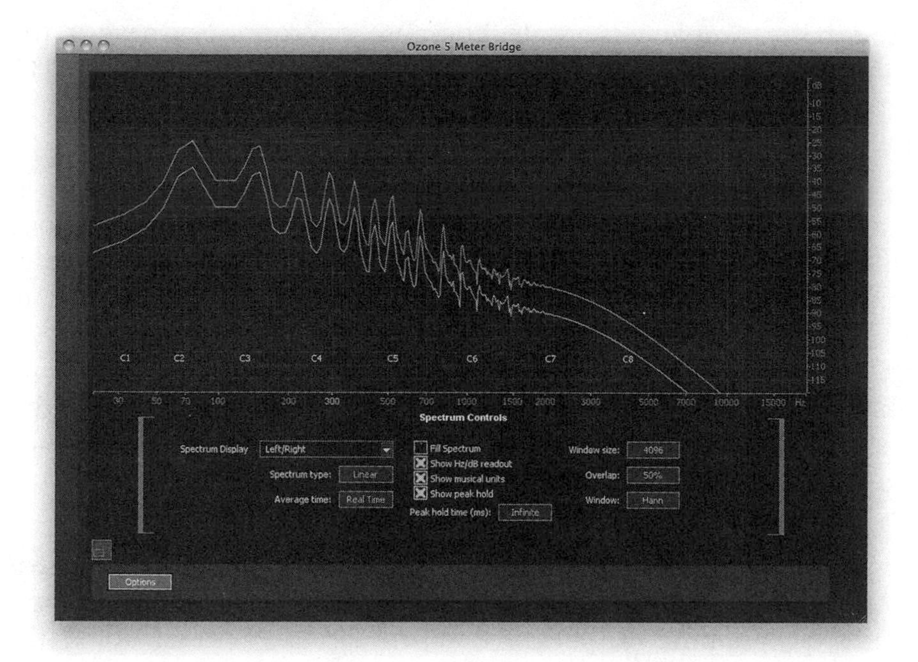

Use FFT analysis to break down the harmonic structure of a given instrument. Often you can articulate an instrument better by accentuating the second harmonic rather than the fundamental.

to modify the balance of an instrument within the entirety of the whole mix, we tend to look initially for the fundamental frequency, as this is the loudest part of the instrument's frequency spectrum. However, you can achieve a similar result by adjusting the relative balance of the second harmonic, which is always twice the frequency of the fundamental and only a few decibels quieter that the root note.

Boosting the second harmonic rather than the fundamental is useful for several reasons. Firstly, you potentially avoid any instruments that might coexist in the same part of the frequency spectrum, which is always an issue with corrective equalization moves that are applied across a master as a whole. Secondly, by being further up the frequency spectrum, you can potentially help an instrument articulate itself over a smaller pair of speakers, which is always an issue for bass instruments with deep fundamental frequencies.

Removing sibilance

Sibilance is problem caused by an unnecessary enhancement of fricative consonants in the human voice – 's' or 't' sounds, in other words. The problem is created because microphones have a frequency peak around the area of sibilance (usually around 4 kHz, depending on the singer), which is often compounded by the sloppy application of EQ.

Ideally, sibilance is best addressed while mixing, when the vocal is distinct and separate from the rest of the track, using a technique called *de-essing*. In this example, a compressor is made 'frequency conscious' by being fed a version of the vocal, via the side-chain input, which has been heavily equalized so that the sibilance is extremely prominent (this usually means a heavy boost around 4 kHz, or so). As the equalized side-chain is being used to drive the compressor's gain reduction circuitry, the compressor can then attenuate the sibilance as and when it occurs, by reducing the output accordingly.

However, in mastering we need to be a touch more imaginative, as the use of a broadband compressor would attenuate the entirety of the programme output so that the sibilance turns down the whole track! When mastering it's best to use some form of *dynamic equalizer*, or a dedicated plug-in such as Sonnox's SuprEsser. As with a conventional de-esser, the dynamic equalizer starts with a

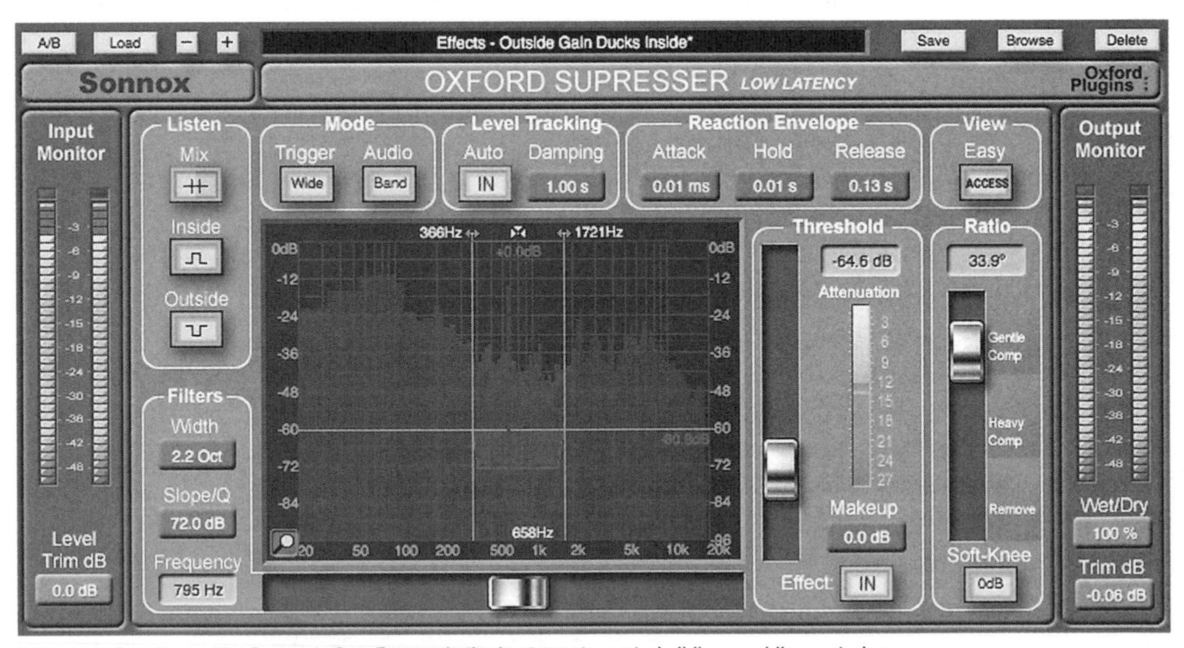

A Dynamic Equalizer – like Sonnox's SuprEsser – is the best way to control sibilance while mastering.

frequency-selective side-chain input, so that it only responds to a given part of the frequency spectrum. Rather than applying broad-band gain reduction, though, the dynamic EQ only attenuates the part of the frequency spectrum defined by two crossover points. This allows the sibilance to be removed while the rest of the music passes untouched.

Beyond sibilance there are plenty of other frequency issues that a dynamic equalizer can be turned to, including a splashy pair of cymbals and plosives in a vocal. They key to the need for dynamic equalization (as apposed to conventional equalization) is that the unwanted frequency artefacts occur sporadically throughout the duration of a song, rather than being a continuous feature. In this case, a dynamic equalizer preserves the general integrity of the track, while still offering the option to keep unruly elements in check.

5.7 Selective equalization – the left/right and M/S dimension

In most situations, equalization is applied to both the right- and left-hand channels of a stereo signal simultaneously. Conceptually speaking, this combined approach makes most operational sense, as the majority of frequency issues aren't localized to one side of the stereo image, and it's often the case that a mastering engineer will want to forge some form of continuity between the left and right channels of the mix to create a stable and balanced master.

Despite the clear benefits of processing the left- and right-hand channels together, there are examples where you want to adopt a more localized solution, so that one side of the mix is equalized differently to the other. Treble boost, for example, can sound sweetening across the majority of the mix, but might push a one-sided hi-hat (mixed slightly to the left- or right-hand side of the mix) in an aggressive way. By unlinking the EQ, the treble boost can be directed more towards the apposing channels (maybe the left channel having a 0.5 dB lift, for example, while the right is boosted by 1.5 dB) allowing you to gain slightly more control with respect to where and how the boost is applied.

On the whole, the use and application of discrete left/right equalization centres on your ability to discern frequency-based issues between the left and right channels. Clearly, if the mix is particularly monaural, there's little or no need to apply equalization this way. However, where you can hear a specific differences between the left and right channels – maybe one or more instruments sitting either side of the sound stage, for example – then the use of unlinked equalization becomes a distinct and interesting proposition.

Unlinking your EQ will allow you to differentiate the application of EQ, maybe applying more top-end boost, for example, to the left-hand side.

Going one stage further with this 'selective EQ' concept, we enter the M/S dimension, creating a different EQ curve for the mid and side components of a stereo mix. We've already touched on M/S processing as part of the previous dynamics chapter, but this is arguably the moment that M/S really gets to shine, as it can perform some significant improvements on a range of different mastering applications.

To best understand the application of M/S equalization, let's take a look at it on a channel-by-channel basis:

Equalizing the side channel

Unusually, we're going to start with the side channel, largely because M/S processing can perform some rather neat tricks in this area. Firstly, let's consider the presentation of bass, which tends to favour a strong mono image, largely as our ears have difficulty discerning stereo information below 100 Hz (in short, it's irrelevant). By rolling off the bass on the side channel (either using a filter or a low shelving curve) we create a natural emphasis towards the mid channel. Of course, the bass is still present and active – we just negate any excessive and unwanted divergence in the extreme 'sides' of the master.

Two ways of performing a similar bias towards 'mid bass' include the Mono Maker feature as part of the Brainworx bx_digital V2 plug-in, and the elliptical filters as part of the Maselec mtc-6 mastering control. Both solutions allow you to refocus the bass, putting stereo signals below 40–360 Hz in the centre of the stereo image. The result is a tighter bass that translates well across a number different speaker systems.

Another interesting side-based equalization move is a simple treble lift around 10–12 kHz. This is particularly valuable given the type of material that tends to sit at the sides of the stereo mix, such as hi-hats, cymbals, double-tracked acoustic guitars, backing vocals,

Rolling-off bass in the side channels can keep the bass-end of your mix tight and focused in the centre of the soundstage.

shakers panned slightly off-centre, and so on. In effect, the sides of a mix are often about musical details – providing an interesting musical texture for the lead instruments to sit in rather than eating up the dominant parts of the musical spectrum. By enhancing the highs on the side channel we effectively increase the vibrancy and life of the stereo image in a way that really counts, more so than if the same colour was applied to the mid channel.

Equalizing the mid channel

Equalizing the mid channel is largely an opportunity to apply a unique EQ curve to instruments sitting in the centre of the mix. For example, we could apply a touch of sibilance-taming equalization to the mid channel without fear that we'll also affect the cymbals and

Equalizing the mid-channel provides access to some of the key instrumentation in the track - like the vocals, kick and the bass – without affecting details in the side channel.

hi-hat, as we would if the same move was applied to a conventional L/R mix. Likewise, a bass boost adds a touch more power to the mid channel, and is particularly effective when combined with a bass cut in the side channels. Finally, also consider a boost in the 1–3 kHz region as means of better articulating the principle lead instrumentation found in the mid channel.

5.8 Subtle colouration tools

Although equalization is the primary tool for modifying the timbre, there are other methods and tools that you can apply to change to tonal colour of your finished master. In many respects, these tools address the principle deficiency of equalization: that it can only modify spectral content which already exists, rather than creating new harmonics. In short, no amount of treble boost will add air to a recording made with a dynamic microphone with a limited HF response. However, by using tools such as *excitement* and *distortion*, we can actively create new harmonic material to enhance the frequency content already present. Equally, tools such as multiband compression can also significantly change the way frequencies are presented, adding a harmonic intensity that equalization can never create.

Rather than starting with some of the more extreme ways of modifying timbre, it's worth highlighting the small but important ways in which a variety of signal processors can change the feel and perception of the mix. As you'd expect, we're talking about very small differences here, but when applied across a mix as a whole, even the untrained ear will notice shifts in the feel and vibe of a finished master.

Non-linearity

All analogue signal processors have a degree of non-linearity to them – in other words, they add distortion. Among other things, distortion changes the harmonic content of a signal, adding additional harmonics which can make a recording appear fuller, brighter or marginally coloured in varying amounts. Even in software, 'modelled' plug-ins often seek to replicate these unique non-linearities, so it might be that even with the plug-in applying marginal amounts of signal processing, it's still transforming the sound of your master.

There are many ways of introducing non-linearity into your mastering signal path, whether you're using software or hardware. Old valve compressors and equalizers often impart a useful colour shift to the material that passes through them, and you can often change the amount of non-linearity depending on how hard you drive the unit in question. Units like the Fairchild 670 compressor, for example, offer some significant differences based on how hard you drive the input and output. Experimenting with these devices across their full dynamic range is important to understanding the variety of effects you can achieve with them.

One of the most interesting non-linearities, though, is the contribution of analogue tape. The precise properties and sonic performance of tape could arguably fill a chapter in itself, and of course there are plenty of variables given the precise tape stock and the calibration of the machine that you use. However, for the purpose of being succinct, the process of 'saturating' tape (in other words, printing a signal so that it's relatively hot) creates a degree of additional harmonic material, as well as a reduction of transient information (a subject that we'll return to in Chapter 6 when we look at loudness). In theory, tape could actually be considered a tool for both timbral and dynamic modification, rather than just a means of adding new harmonics.

Tape saturation – either in plug-in or hardware form – adds some useful spectral and dynamics changes to musical material.

Spectrally speaking, tape saturation has some interesting properties worth noting. Firstly, the added harmonics tend to have a bias towards odd-ordered harmonics, often providing a flattering reinforcement and weight to the low-end of the mix, whether it's a bass guitar or a kick drum. Depending on the condition of the tape, the speed it's running at (usually 15 or 30 ips) and any pre-emphasis equalization, there'll also be some interesting developments in the high-end. Although the saturation creates additional high-end harmonics, there's also a degree of high-frequency roll-off, which is one of the key contributors to the 'warming' effect we often attribute to tape. Beyond saturation, there are also effects like wow and flutter, which add small pitch fluctuations that, in controlled amounts, add another layer of 'warmth' to the proceedings.

For the purposes of mastering, therefore, some engineers may choose to print a mix that has been supplied to them in a digital format onto reel-to-reel tape, almost using the tape machine as another form of signal processor. In theory, the tape negates the sterile 'thin' quality of sound that can be attributed to music produced entirely in the digital domain, even with aspects such as wow and flutter arguably contributing to the musicality of the end result. Of course, not all forms of music benefit from this treatment (detailed classical recordings and hard-edged electronica both spring to mind), but it is a useful option to have to hand. Even if you haven't got access to a reel-to-reel machine there are plenty of software alternatives, and hardware emulators such as the Portico 5042, that can replicate the unique sonic characteristics of tape.

The 'Sound' of components

Although this point is somewhat linked to the last comments about non-linearity, it's important to note that all analogue components deliver their own unique sonic quirks – whether it's the valves of a variable-MU compressor, the FETs inside an 1176, or the hand-wound transformers used in a Neve equalizer. All of these components change the colour of the sound, which can be a positive or detrimental addition to the sound your trying to produce.

Phase shifts

As we have discussed, all analogue equalizers introduce phase shifts whenever they apply a cut or boost. The 'sound' of your 10 kHz shelving boost is as often as much about the phase shifts that it induces as any modification to the frequency content. As you'd expect, different equalizers will induce different phase shifts, and some engineers will actively negate these colouring effects by using a phase-linear equalizer.

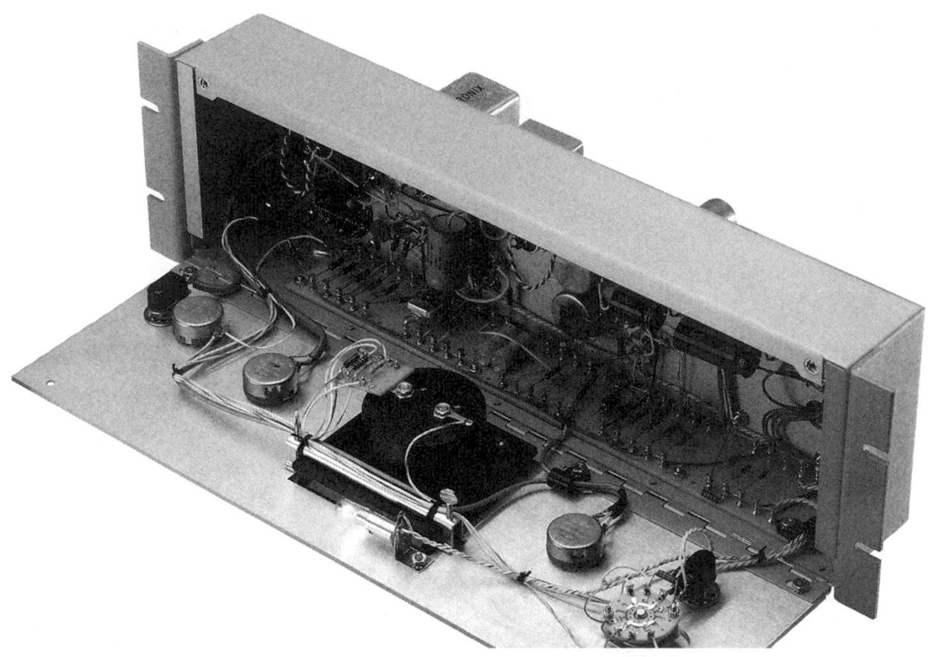

FETs, Transformers and Valves can all colour the sound in interesting and musically useful ways.

Converters

Although converters have become increasingly neutral in their sound, it's not uncommon for mastering engineers to run the signal in and out of the analogue domain partly to capture the particular sound of the converter. Older Apogee designs, for example, exhibited some interesting colours in the mid range that changed the drive and focus of the music that passed through the converters. In the quest for increasing loudness, some engineers have even gone so far as to deliberately overdrive the converters for the unique signal distortion the signal clipping delivers. Whether it's analogue or digital, every device has a 'sound' to offer.

5.9 Extreme colour – the multiband compressor

We dealt with multiband compression in detail as part of the last chapter on dynamic control, but in many ways it's worth remembering that a multiband compressor can change the timbre of a recording as much as, if not more than, an equalizer. In some situations, therefore, it might be pertinent to question whether changes to the audio spectrum should be carried out with an equalizer or a

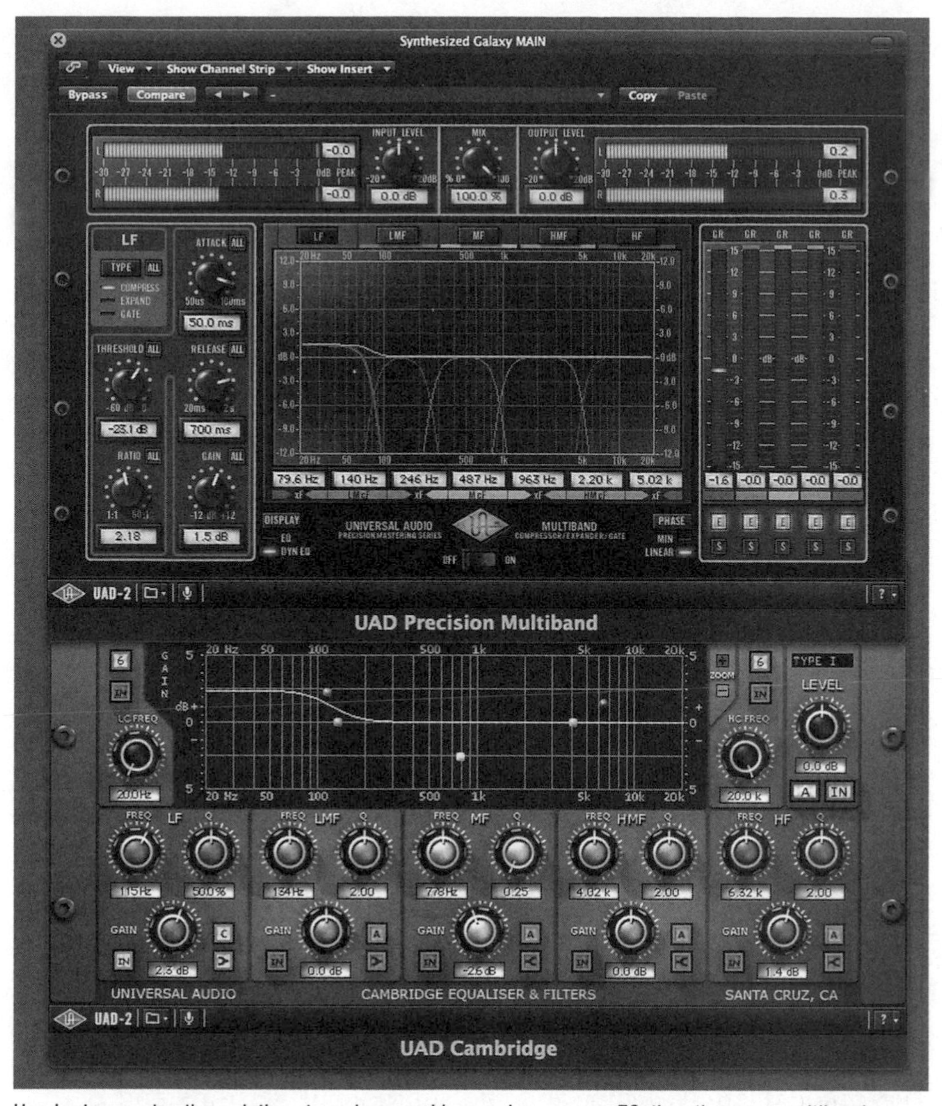

Here're two contrasting solutions to an improved low-end – one uses EQ, the other uses multiband compression.

multiband compressor. A weak bass, for example, could be rectified by a boost at 100 Hz, or by compressing the bass spectrum slightly harder than the rest of the mix, as well as raising its output level slightly as part of the gain makeup.

Ultimately the decision as to whether you use equalization or multiband compression to solve a timbral problem largely falls to down to whether there are dynamic issues involved as well. In the previous example of a weak bass, you need to ask whether the low-end is inhibited through dynamic inconsistencies, or simply by the bass levels being slightly too low in the mix. By varying the amount of compression and the eventual gain makeup level, multi-band compression can change both the level and intensity of a given frequency band. However, if the same result can be achieved simply through a touch of EQ, it's much better for the track's musicality to use the 'less is more' approach.

5.10 Exciters and Enhancers

Throughout the history of recording, mixing and mastering, a number of devices have come along to usher in some magical way of manipulating frequency content, either by adding new harmonics, or playing with existing harmonics to change their perceived balance.

Aphex Aural Exciter

Aphex's original Aural Exciter was arguably the first 'mojo' enhancer to be brought to market back in the mid-1970s, and was initially hired out on a 'cost per minute of recorded music' basis. It soon became a desirable and elusive sonic weapon, capable of adding the sheen and sparkle back into music. To put this into context, the 1970s was also an era when treble itself was an elusive quality, with a degree of generation loss from working on tape. In theory, the Aural Exciter added back the 'air' that was lost as the master tapes became increasingly worn – something that seemed impossible to recreate using EQ.

The original Aural Exciter and the subsequent versions and derivatives that followed it all worked on the same principle – that of 'filtered' distortion. The first stage is to put the input signal through a high-pass filter, negating the risk of adding additional harmonic information across the entirety of the mix. This filtered version of the input is then marginally distorted, creating additional harmonic material. The final stage is to blend some of this 'filtered distortion' back into the signal path, adding a subtle amount of sparkle in the sound spectrum above 10 kHz or so.

Aural Exciters are a useful way of adding sheen and air in situations where the original source material lacks treble.

As with the original process, aural *exciters* (or *enhancers*) work best when the material is deficient in the upper portion of the frequency spectrum – usually when a suitable shelving boost doesn't give the desired 'air' lift you'd expect. Their application can become addictive though, particularly as the ear can soon get used to the added frequency material they generate. Exercise some caution when you apply any form of exciter or enhancer, and make sure you reference some trusted recordings to make sure you're not overcooking the mix of the 'excited' signal.

Waves Maxx Bass

There are a number of interesting bass-enhancement technologies available, but Waves' Maxx Bass has arguably proved itself to be the most popular. In some respects, it goes back to some of the concepts we talked about with respect to equalizing the fundamental and first harmonic. In the case of Maxx Bass, new harmonics are generated based on the original fundamental frequencies contained in the subsonic area of the mix. In theory, these reinforced second and third harmonics increase the presence of the bass, allowing you to attenuate the level of the fundamental and therefore articulate 'deep' bass on speakers that haven't got a particularly extended bass response.

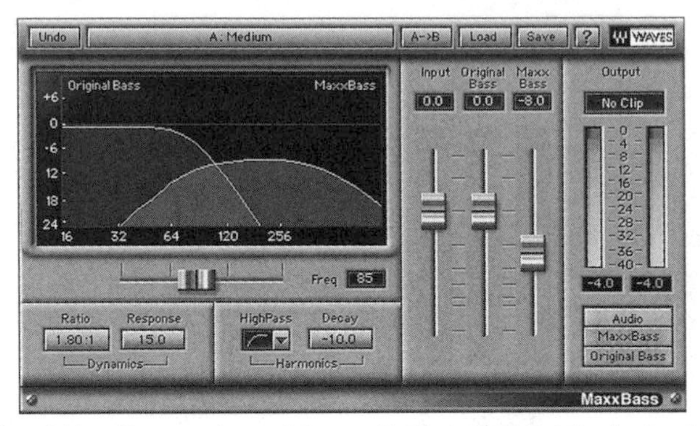

Technology like Waves' Maxx Bass can be useful way of better articulating the low-end over playback systems with a limited frequency response.

CHAPTER 6

Creating and Managing Loudness

In this chapter

6.1 Introduction: A lust for loudness 134
6.2 Dynamic range in the real world 135
6.3 All things equal? The principles of loudness perception 136
6.4 Loudness, duration and transients 138
6.5 Loudness and frequency 140
6.6 The excitement theory 143
6.7 The law of diminishing returns 145
6.8 Practical loudness Part 1: What you can do before mastering 147
Simple, strong productions always master louder 147
Don't kill the dynamics – shape them! 148
Understand the power of mono 148
Control your bass 149
6.9 Practical loudness Part 2: Controlling dynamics before you limit them 149
Classic mastering mistakes No. 1: The 'Over-Limited' chorus 150
Moderating the overall song dynamic 151
Moderating transients with analogue peak limiting 152
Controlling the frequency balance with multiband compression 153
6.10 Practical loudness Part 3: Equalizing for loudness 154
Rolling Off the sub 154
The 'smiling' EQ 155
Controlling excessive mids 156
Harmonic balancing 156
6.11 Practical loudness Part 4: Stereo width and loudness 158
Compress the mid channel 159
Attenuate bass in the side channel 159
6.12 Practical loudness Part 5: The brick-wall limiter 159
How much limiting can I add? 159

Knowledgebase 161
Setting the output levels 162
Multiband limiting 163
6.13 The secret tools of loudness – inflation, excitement and distortion . 163
Sonnox Inflator 165
Slate Digital Virtual Tape Machines 166
Universal Audio Precision Maximizer 167

6.1 Introduction: A lust for loudness

For most engineers and musicians taking their first steps into mastering, the attainment of 'loudness' is often one of their principle objectives. Certainly, given the apparent differences between an unmastered mix and the extreme loudness found on many commercial CDs, it's easy to see the allure of loudness and why it's often the first port of call. However, as many people soon discover, loudness is a considerably more elusive entity than most of us initially imagine, often requiring more than just a single 'magic bullet' plug-in to achieve.

In truth, loudness is one of the most intriguing components of the mastering equation – capable of producing transformative results (a controlled amount of loudness benefits almost any mix), but equally it is a tool that can ruin an otherwise good recording, even making it unlistenable! What is apparent, though, is that few of us can afford to ignore it, especially with an increasing amount of music being subjected to almost painful amounts of limiting!

In this chapter, we're going to take a detailed look at the concept of loudness, and gain an understanding of the important difference between a sound meter's response to amplitude and the radically different ways our ears perceive the relative loudness of a given sound. As we'll see, it's not a straightforward equation, and it soon becomes apparent why loudness is such an elusive entity, and why many struggle endlessly in search of the perfect way of creating a 'super loud' commercial-quality master. As a balancing note, we'll also examine the importance of dynamic range, and how a more measured approach to loudness might actually make your music sound better.

Moving on from the concepts of loudness, we'll also explore a series of practical steps that you can take to ensure that your masters sound as loud as possible. Starting off from the production itself, we'll look at practical tips that the professionals use to guarantee maximum loudness potential. Moving on to the mastering process itself, we'll also look at a variety of ways in which loudness can be injected into a master – not just using conventional tools such as limiters, but also a broader investigation into how a range of signal processors all have an impact on the eventual loudness you can achieve.

6.2 Dynamic range in the real world

Before we delve into the specifics of loudness, we'd like to take a step back and consider the concept of *dynamic range*, or, more specifically, our perception of signals that occur across a wide range of amplitudes. For some, dynamic range could be considered the antithesis of loudness – the very thing that a string of compressors and limiters attempts to remove as part of creating loudness – but in reality it's an important concept to understand, and the origination of everything our ears respond to.

In the real world, our ears can perceive sounds across a tremendously varied range of amplitudes – from the infinitesimally small sound of a grasshopper rubbing its legs together, to the altogether more decibel-heavy sound of a thunderclap or an explosion. Interestingly, this ability to perceive sounds over such a wide dynamic range (around 130 dB, to give it a number) is partly explained by the ear's ability to protect itself in response to loud sounds, arguably creating a natural form of 'auditory compression'. Indeed, anyone who's experienced a loud rock concert will be well aware of the ear's natural defence mechanism, and will have experienced the effect that it achieves (something which we'll return to later on!).

Arguably, the big shift in our relationship with dynamic range came with the invention of electronic sound recording technology. In short, recording technology has somewhat struggled to compete with the ear's ability to work over a wide dynamic range. As a starting point, consider the power of an amplifier and speaker needing to match the same dynamic range as small chamber orchestra. While it's conceivable that a home hi-fi could possibly match the acoustic power of a few woodwinds, you'd need something much closer to a full-blown PA once the brass and percussion start playing! Equally, old tape-based recording systems often struggled with poor noise performance and distortion – either being too hissy on quiet passages, or too distorted once the full orchestra started playing.

The full dynamic range of acoustic music is something that the electronic domain has always struggled to capture.

What is apparent, therefore, is that sound in the electrical domain struggles to compete with the dynamic range of the real world. Even though recording technology has developed (certainly, high-resolution digital recording comes much closer to our ear's dynamic range), it's still surprising to note the general inferiority of playback systems, especially with respect to the amount of music now being consumed via a set of small speakers attached to a computer. Ultimately we're all trying to squeeze as much information as possible through a medium where dynamic range, and crystal-clear sonic reproduction, doesn't come in large amounts!

Given the inherent problems of dynamic range in recorded music, and the fact that necessity is often the mother of invention, it's no surprise that the deliberate abuse of dynamics has embedded itself into music production. From the first day a compressor entered the studio, engineers soon realized that the 'distortion' of dynamic range, though technically an imperfection, actually led to their records sounding more exciting and vibrant to the listener. In truth, therefore, although we like to think of the loudness war as being a relatively recent phenomenon, the seeds were sown way before aggressive CD mastering started taking place.

6.3 All things equal? The principles of loudness perception

To fully manipulate loudness, it's vital to understand the concepts behind this unique part of sound recording practice and auditory perception. Unlike the majority of the production process, which largely deals with the objective measurement of frequency and amplitude using frequency analyzers and decibel meters, loudness is an area principally defined by our psychological response to a piece of music. Put in simple terms, *amplitude* is a term used in measuring the physically intensity of *sound pressure levels* (or SPL, for short), whereas *loudness* specifically relates to our psychological response to a perceived amplitude – in other words, how loud we think it is!

The electrical measurement of amplitude and our perceived understanding of loudness are often radically different.

Given its psychological nature, therefore, loudness is an incredibly subjective measure, and even though we understand some of the overarching principles, it's clear that our relationship with loudness is far more complex than any of us will ever fully understand. What is clear, though, is that there can be some startling differences between the electrical amplitude of a signal and its percieved loudness, which explains why a commercial CD might sound 10 dB louder than a mix that you've just finished in your studio, even though both audio files peak at 0dBFS. Mastering engineers implicitly understand the phenomenon and power of loudness, and know how best to utilize it as part of maximizing the potential of the music they work with.

In trying to understand loudness, what soon becomes apparent is that it's born from the interaction of a number of parameters. The ear is a complex mechanism, as are our brains, and to think that loudness is purely created by a few clever tricks with amplitude (using brick-wall limiters, in other words) is somewhat missing the point.

6.4 Loudness, duration and transients

One of the defining qualities of loudness perception is the duration of a sound. Short sounds that might meter full-scale (0dBFS, in other words) can seem relatively quiet, whereas a signal that has a consistent sustained dynamic, even somewhere below full-scale meter readings, can be perceived as being significantly louder. This is easy to hear for yourself by comparing a normalized drum loop created on a drum machine (ideally without any compression), and a simple sine wave also normalized to the same figure. Try pulling down the fader on the sine wave until you feel that both sounds have an equal loudness – depending on the drum source, you might need to attenuate the sine wave by as much as 12 dB.

These two audio tracks have been balanced so that they appear to be equally loud. Notice how the pure sine wave has been attenuated by 12dB!

Understanding that duration plays an important part in loudness perception opens up some interesting observations about music in the electronic domain, and in turn, some of the principle techniques used to control and create loudness. A significant barrier to achieving loudness is transient energy – a short burst of sound energy that kick-starts a range of percussive sounds, or instruments with a defined attack – usually created through strings or drum skins being picked, plucked or hit. Transients, of course, give music drive and bite (contrast soft, muted strings against a drum kit, for example), but they also contain a large proportion of sound energy with in a given signal. Indeed, look at any uncompressed drum recording and you will see a series of sharp transients dictating the peak levels reached in the given recording.

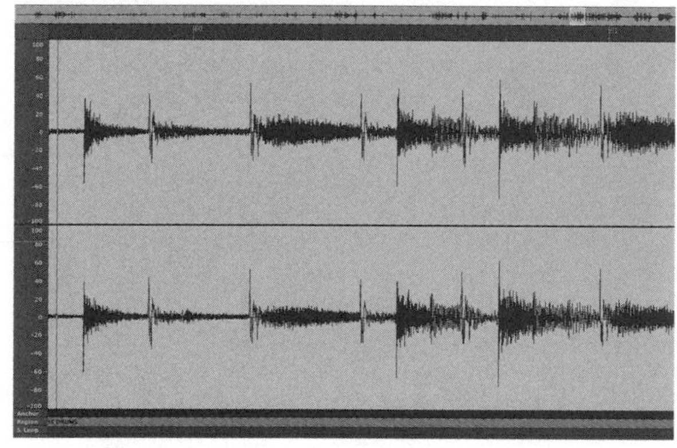

Transients are an important part of the vitality and energy of music, but they can also inhibit the amount of loudness you achieve.

As such, some form of transient control and reduction is an important part of creating loudness. Looking first at transient control, you can see that transients can be quite erratic in their level – our ears tend to be very forgiving in this regard, but the precision of a digital meter, and indeed, digital recording itself, isn't so tolerant. If you've ever recorded a live drum kit, for example, you'll be well aware of the need to leave 10 dB or so of headroom, only to find that the meters 'hit red' when the drummer wallops the snare ever so slightly louder on the first beat of the final chorus!

Interestingly, analogue tape is notoriously forgiving on transient peaks, applying small amounts of tape saturation (or distortion) on the occasional stray hit, and thereby keeping the transient levels in some degree of order. Of course, in an entirely digital production chain, it's highly likely that these transient peaks have passed through the majority of recording and mixing without any of these pleasant 'non-linearities' slipping in, and therefore the average level is several decibels quieter than a comparable analogue recording. While this preservation of transient detail shouldn't necessarily be seen as a 'fault' of digital recording, it does explain why transient control, and indeed, the role of the limiter, is so important for mastering today.

Tape has long been favoured for its 'forgiving' approach to loud transients, being quick to react and effective at maintaining a more consistent averaged level.

Moving beyond the concept of transient control, we then start to look at the technique of transient reduction, whereby the transient energy is deliberately and significantly reduced to increase the average level of the master, and hence it's perceived loudness. As you'd expect, reducing transient energy can have a significant impact on the loudness you achieve, although you also start to compromise some of the detail, bite and percussive detail of the track you're trying to process. Think of aggressive transient reduction as a form of 'easy loudness' – a weapon of last resort once you've explored all other avenues.

6.5 Loudness and frequency

One of the biggest misconceptions with loudness is that it purely relates to amplitude. Indeed, given our obsession with peak digital meters, and the need for a final mix to be regularly hitting 0dBFS, you can see why we've largely ignored the importance of timbre in relation to loudness. In reality, though, we're missing a significant trick, and we would argue that most of the 'loudness' associated with a commercial CD is as much to do with its timbre as the amount (or choice of) limiting applied during the mastering process.

Indeed, you only need to 'level match' your original audio file and the commercial master (this usually involves attenuating the commercial master by around −6 dB) to realize how the timbre is just as different as any perceived increase in the track's level.

As with amplitude – where duration is as important as sound pressure levels – our relationship with frequency is complex and difficult to understand precisely. First off, we have a decidedly non-linear response to frequency – in that certain frequencies appear louder to our ears than others. This non-linear frequency response is partially explained by our ears being particularly sensitive and developed around the frequencies of speech, allowing us to hear frequencies around 2–5 kHz particularly well. Above 5 kHz, and below 100 Hz or so, our frequency response rapidly drops off, meaning that our ears are much less sensitive to these sounds.

To add even more confusion into the frequency equation, the precise nature of the non-linear response to frequency changes with the overall amplitude. In short, our frequency perception becomes flatter as amplitude rises, so that we hear more bass and an increasingly vibrant treble as the volume of playback is lifted. If we want to fool the ear into thinking that it's hearing a loud mix, we need to ensure that both treble and bass are clear and defined, and 'hyped' to some extent in comparison to the real world. Indeed, following this principle, a large amount of hi-fis produced in the 1970 s and 1980 s had

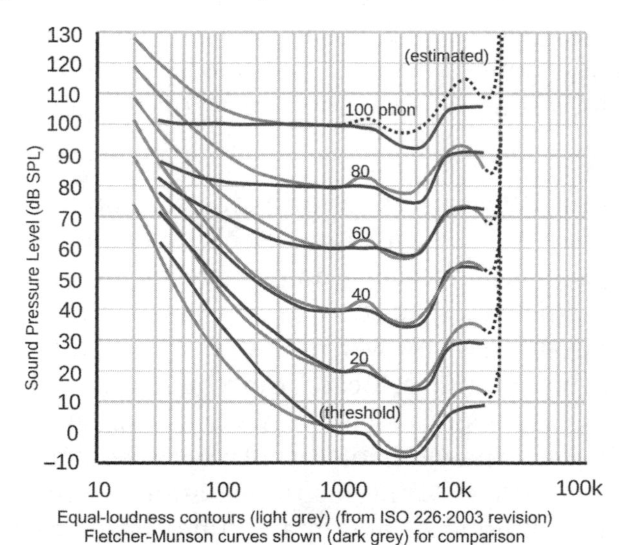

Fletcher and Munson's equal loudness curves prove that our perception of frequency changes with amplitude. In short, loud music appears to have a more vibrant treble and bass response.

so-called 'loudness' buttons, designed to be activated for 'quiet' listening, although in actuality most listeners incorrectly kept loudness engaged all the time!

As well as looking at the enhancement of bass and treble to fake loudness, we also need to consider the role and impact of the mid-range. Where an instrument or mix is particularly prominent in its mid range (where the ear is most sensitive) we create a master that is aggressively or unpleasantly loud – our response isn't linked to excitement, but instead urges a listener to turn down the volume control. A good example of this is something like a harpsichord, which tends to have a particularly prominent biting mid-range and not a great deal of dynamic range. In the case of mastering a solo harpsichord recording, therefore, the need to apply any techniques to create loudness should be considered a largely pointless exercise! In short, your approach has to be adaptive, and can't just rely on a one-size-fits-all solution.

The final part of the frequency equation in respect to loudness is the use of sound energy. Put simply, bass-end requires far more sound energy to be effectively conveyed than the rest of the audio spectrum – a woofer, for example, needs more electrical energy behind it than a tweeter. You can also see the effect of bass and sound energy on a console's meters when you're mixing. Bring up the bass and kick channels and notice how much movement they generate, almost hitting full-scale throughout the majority of the track. Now balance the other instruments and notice how little additional energy is added to this initial level, indicating just how much power is contained in the low-end of the mix.

The relationship between bass and loudness is, therefore, a complicated one. As we've seen with the Fletcher and Munson curves, the presence of bass is an important part of perceiving loudness – in other words, if a track is bass-light it won't appear to be particularly loud. Conversely, it's also easy to over-cook the amount of bass, which in turn restricts the amount of headroom you're left with for the rest of the mix. In that respect, a bass-heavy track can also appear quiet (especially if the bandwidth of the playback system is restricted), largely because the rest of the instrumentation is unnaturally quiet in the overall mix. Indeed, how many times has a record sounded great in a club, but translated poorly to the TV or the Radio, where the bass isn't fully present?

6.6 The excitement theory

Beyond the basics of amplitude and frequency, there are many factors that seem to affect our response to music and the perceived loudness of a master. Indeed, when you're writing or recording music there's often a tendency to push the control-room levels up, finding a near exact point where the music seems to 'come alive' and fill the room. Arguably this approach makes sense, as we're trying to inspire and excite our creative synapses in much the same way as we might feel inspired by seeing some live music played loudly, or by a powerful score played over a cinema's sound system. Fundamentally, it seems, we have a desire to 'move air' through the creation of music.

When it comes to mixing and mastering, though, we need to temper excitement with objectivity. Some of the most well-respected music engineers in the business – including Chris Lord Alge and Bob Clearmountain – have spoken about how they like to work at relatively modest monitoring levels. By working at such a conservative level they feel that they can gain a more objective view of the track – maybe it's less exciting than when the monitors are cranked up, but it's more representative of how the majority of listeners will first encounter a piece of music. Another significant benefit is the ability to work for long durations of time without ear fatigue, allowing you to make informed decisions throughout a whole day of work, rather than just in the first hour. In short, if you can make quiet music sound good, it'll sound great however loud it's played!

For many musicians working on mastering their own music, it's also worth considering how objective we are with respect to assessing our eventual loudness. Although we have no scientific basis for this assertion, we'd argue that 'ownership' of the music often distorts your perception of its loudness by several decibels, making it difficult to objectively reference it against other recordings. To a large extent, it's possibly driven by a need to make your music sound better – or louder – than the competition, but in truth, the subjective differences are probably born as much from musical differences as any semi-objective reading of its loudness. It's easy to see, therefore,

Loudness metering

The loudness wars have been around for a very long time and, we dare say, have dominated much of the text in this book. However, it is encouraging to note that there are now significant moves afoot to rebalance the disturbance in the loudness force.

Many, starting with perhaps Bob Katz's AES conference paper, have been rallying for years for the audio (and music) industry to 'aim' for a level of loudness that is lower than the absolute maximum available. This, in theory, might mean that we aim for a level of RMS loudness lower than the 0 dBu peak level. This would permit dynamics and restore necessary headroom. For more details of Katz's argument and an excellent opening description of the metering issues, look at his paper here: www.aes.org/technical/documentDownloads.cfm?docID=65.

Part of what Katz is calling for in his paper has been the basis for some new standards. The premise is that the overall, mean, loudness of a track would be measured rather than the peak. For broadcast this is a standard more likely to be adopted as there's less of a race for the loudest programme around (unless it's the commercials). The loudness wars are the issue for us. The 0 dBu level LED has become a target rather than a warning beacon. This results in a skewed perception that flat-lining our dynamics makes music louder.

In the past few years significant progress has been made in this area, including the first new loudness metering standard provided by the International Telecommunication Union's (ITU) 1770-2 specification. The second of importance is the subsequent metering standard laid down by the European Broadcast Union (EBU) in standard R-128.

Combined, these two standards together introduce a new loudness metering standard measured over time, and encourage users to seek out both *true peak* levels and also a new numeric value called *loudness range* (LU). The concept behind these new standards is to perceive, and now be able to measure accurately, the true loudness of material. While being overtly called for from within the broadcast industry, this sort of concept is something that many in the music world, including Bob Katz, have been calling for us to address for a long time. That time might be just around the corner.

If the music world were to adopt the R-128 standard and aim for a mean loudness, we would be able to retain dynamics and excitement in our music again. Therefore it is really welcoming to find that the standard is trickling into popular software solutions such as Ozone's Meter Bridge where, the R-128 standard is complied with.

On the professional broadcast side, there are a number of manufacturers who provide hardware units for measuring audio that embrace the new loudness standards. One such example is the RTW TM3, which is a small desktop unit with an excellent display showing many different forms of meter.

Understanding the tools you have at your disposal is obviously of central importance. In the current climate, it would be surprising if you didn't wish to, or weren't asked to, contribute to the loudness

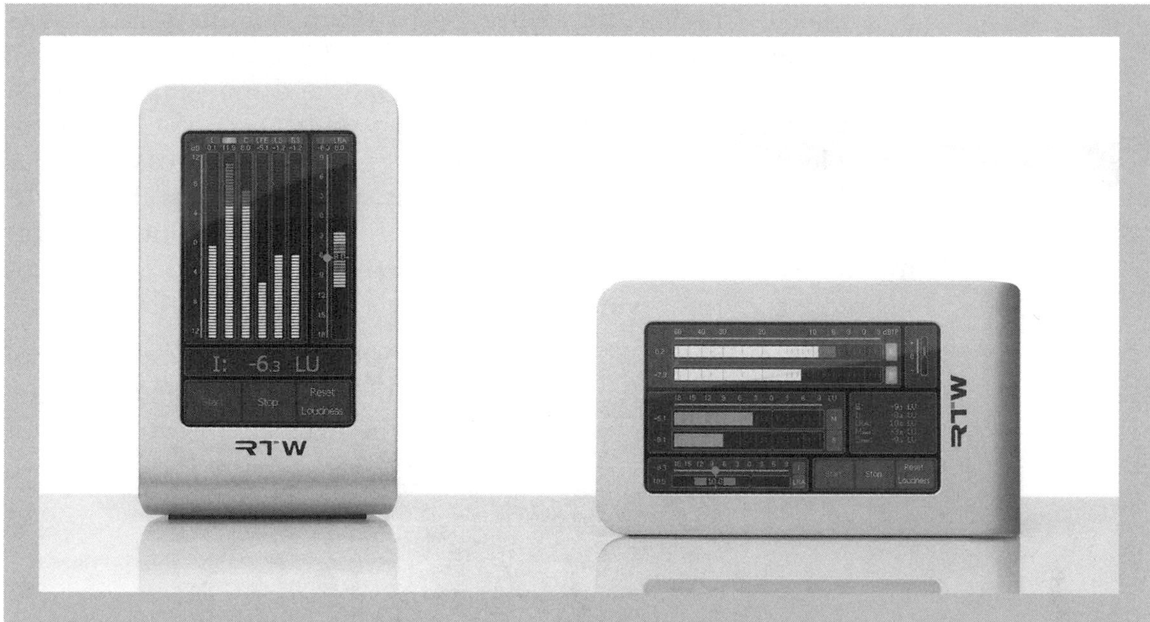

RTW's TM3 is a good example of a recent solution to permit accurate metering of loudness, possessing all the current standards (especially R-128).

wars in some way or other. Clients, or fellow musicians, will want their material to be as loud as the next person, continuing the vicious cycle and keeping things too loud. However, understanding the difference between your peak and RMS levels will mean that you can see your loudness and allow you to start making assessments. If you're fortunate enough to have Ozone or other software that now displays the latest loudness standards, then embrace this, as this may become the norm for our work in the future.

why a 'loudness spiral' is created, with each artist wanting to push themselves slightly harder than the competition!

If you're mastering your own music, therefore, it's worth applying some degree of caution to any loudness that you create. Use objective measurements of loudness (such as a dedicated loudness meter) as well as you own ears, or use a trusted friend's objective viewpoint to see how loud they feel your master sounds.

6.7 The law of diminishing returns

Before we delve into a specific techniques and practices in relation to loudness, it's worth making a cautionary note about the diminishing marginal returns that they offer. To be blunt, loudness is a somewhat finite entity, and a mastering engineer has to strike a balance between what can realistically be achieved, and the point at which the music really begins to suffer. Take the application of a limiter, for example. At first, the initial few decibels produce great results – adding plenty of loudness without the distortion being too noticeable. However, as you push the limiter harder you start to notice that: (a) the mix isn't getting much louder, and (b) the distortion is becoming more and more evident.

With any tool that you apply for loudness, therefore, it's important to evaluate or identify the point at which you're not getting a significant amount of loudness back for the decibels of extra drive or

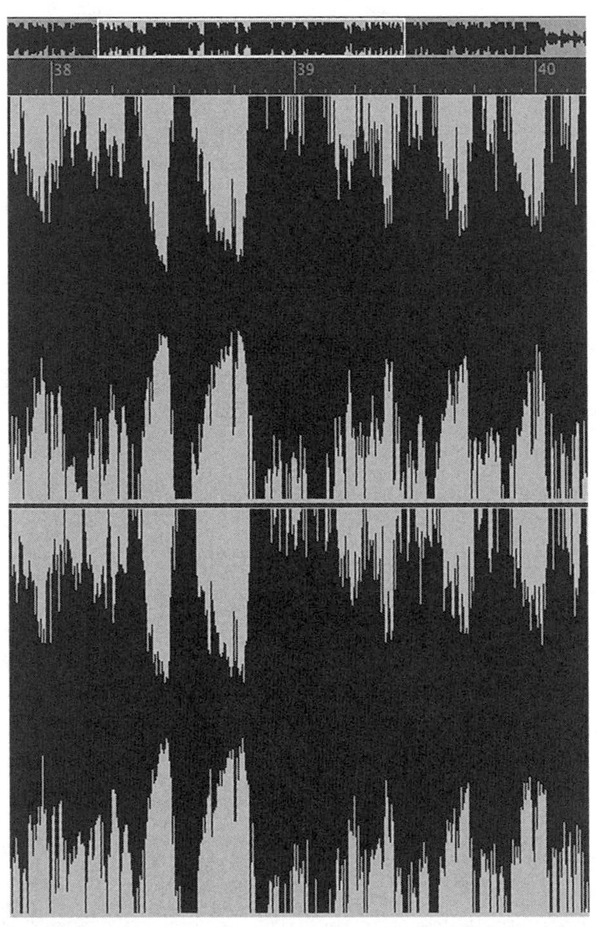

Applying increasing amount of limiting simply adds more distortion rather than an increased sense of loudness.

compression you're adding. Of course, this is a somewhat subjective analysis – somebody listening on a pair of small computer speakers probably won't notice a small percentage increase in distortion, whereas even the smallest increase in distortion will be evident on some well-tuned mastering monitors. Even so, this balancing act is one of the most significant skills for a good mastering engineer.

Another point to remember, especially given the multifaceted nature loudness, is that you shouldn't rely on one device in your signal path being the only means of loudness creation. The best approach is one that is holistic – understanding how a number of different devices (whether it's an equalizer, valve compressor or digital brick-wall limiter) all contribute to the overall loudness that you achieve. The end result is often far more powerful when multiple techniques are used, creating a large difference from a lot of small changes, rather than one device having a heavy-handed and clumsy effect on the results. In short, loudness is a way of being, rather than just a single plug-in in your signal path!

6.8 Practical loudness Part 1: What you can do before mastering

It's interesting to note how many of the problems faced in mastering can be most effectively addressed by better practice before you go anywhere near a mastering plug-in or signal processor. This is never truer than with the concept of loudness. Investment made earlier on in the production process will ultimately allow the tools of mastering – whether it's a limiter, compressor, equalizer or any other form of audio wizardry – to carry out their work more effectively.

Simple, strong productions always master louder

While it's easy to get distracted by multiple tracks, dense sound-layering and complicated arrangements, it's interesting to note that some of the loudest commercial records are often surprisingly simple and direct. Making your productions stand out with a few well-honed sounds will always produce far better results than those hiding behind an ill-defined wall of sound. Simplify the arrangement, prune out superfluous details, and make each sound contribute to the mix in a unique way.

Understanding why the rule of 'simple, strong productions' applies is an interesting insight. On a rudimentary level, running hundreds of tracks often means that their contribution is increasingly marginal. For example, note how the channel fader's average level falls as the track-count rises (or, at least, the stereo buss fader comes down). Looking at the mastering process itself, and you will also notice the signal processors having a more direct and tangible impact on the music. An equalizer, for example, starts to work with specific instruments rather than a broad mush of sound around a given frequency. Equally, a compressor can really start to shape the dynamic structure of key instruments in the mix, rather than squashing a broad swathe of sound.

Don't kill the dynamics – shape them!

Creating a finished mix shouldn't be about squeezing every last decibel of dynamic range from the music. Instead, you should ensure the music is conveyed as effectively as possible. Use compression, therefore, to aid the way in which different parts of the mix sit together, rather than removing dynamic range in an arbitrary way. Interestingly, the use of compression in mixing doesn't necessarily lead to a mix without dynamics, shape and form. Ultimately, compression should shape the blend of sound, rather than fully define the dynamic range of the end master.

When it comes to buss compression, use it for small amounts of sectional 'glue' and avoid letting too much loose on the main stereo buss. All too often, there's a tendency for the mix engineer to start addressing some of the concerns of mastering across the stereo buss. At best, this needlessly doubles-up on certain mastering tasks, and at worst it can create problems that a mastering engineer needs to rectify before they start the process of mastering. Ultimately, it's always easier to add another stage of limiting in mastering than it is to try and remove excessive buss processing.

Understand the power of mono

Although there are techniques a mastering engineer can use to control stereo width, it's worth remembering how the use of stereo can affect the loudness of a mix before it's been mastered. In short, a mono mix is more powerful than a stereo one, although of course, it's also a considerably more one-dimensional experience. The trick here is to rationalize your use of the stereo sound stage – using the centre of the image to convey the power of the track (vocals, bass, kick, snare and so on) while the sides convey detail, interest and dimensionality. If you want the master to be really loud, though, consider pulling as much into the centre as possible, using the extremes just for the occasional effect.

Control your bass

Given the amount of sound energy used to convey bass, ensure that you make full and effective use of the bass-end of your mix. In short, mixing bass 'loud' on a pair of powerful studio monitors sounds great, but you'll only be eating-up the available dynamic range, making the whole mix proportionately quieter than one with a selective amount of bass. Of course, mastering can help rectify many of these problems, but it helps to start establishing some ground rules as soon as you begin to piece the mix together.

One area that's well worth addressing in the mix is any unnecessary overlaps in the bass spectrum, particularly evident in a muddy low-end without any clear bass presence. Use plenty of high-pass filtering on individual channels therefore – either to address issues such as proximity effect, or simply to keep low-mid instruments from straying too easily into the bottom of a mix. Ultimately, there's a lot to be said for leaving just two instruments in the bottom of your mix – kick and bass. Even between these two instruments, it's worth defining which is the deeper of the two, applying a 20–40 Hz high-pass filter to one of the two instruments to define a clear relationship between either sound.

6.9 Practical loudness Part 2: Controlling dynamics before you limit them

As we saw in Chapter 5, there are variety of ways in which mastering can engineer can shape the dynamics of a track using compression. Controlling dynamics, is more to do with the energy, glue and intensity of a master, but you'll also gain an important advantage in respect to a moderate reduction in dynamic range. The benefit here is that the limiter has less work to do, as well as being presented with a version of the track that already has some subjective improvements in its dynamic structure.

The best example of the need from dynamic control, though, is an example of what can go wrong when you extensively rely on a limiter.

Classic mastering mistakes No. 1: The 'Over-Limited' chorus

Since the adoption of aggressive digital limiting, many tracks have exhibited a unique form of 'dynamic distortion' that severely skews the relationship between the verse and chorus of a song. The problem stems from the dynamic differences that are implicit to the musical delivery of a verse and a chorus. On the whole, the verse is restrained, with space to define the narrative message of the song. When the chorus is reached, the energy of the music kicks in, and there's a clear shift towards the emotional content of the song – whether falling in love, falling out of love, anger, loss or happiness! As you'd expect, such different styles of delivery create some significant shifts in the dynamic structure, but what happens when this is aggressively passed through a brick-wall limiter?

At first, the experience of listening to the verse is one that is musically satisfying. Despite the delivery being relatively reserved, the track is hitting full-scale peaks throughout its duration. Although some limiting is being applied, it's only on the occasional transient, so you still get to the feel the energy and life of the piece of music and, most importantly, it sounds loud!

However, when the chorus enters there's a distinct change in the listening experience. Despite the clear change in the way the music is played, there's no immediate difference in the overall amplitude – the music simply sounds more squashed. As a result, the chorus feels more like a 'wall of sound' rather than the first emotional climax of the song – the drums and percussion lose their bite, background details suddenly become overly enhanced, and the sound starts to become distorted. Ultimately, the sound becomes so painful that you reach for the volume control.

Although this example is slightly exaggerated, it clearly illustrates how the wrong choice in loudness creation can ultimately lead to the original musicality of the track being compromised. Of course, the ultimate flaw is the need to turn down the volume in response to the sound of the chorus, which is the completion of a self-defeating circle.

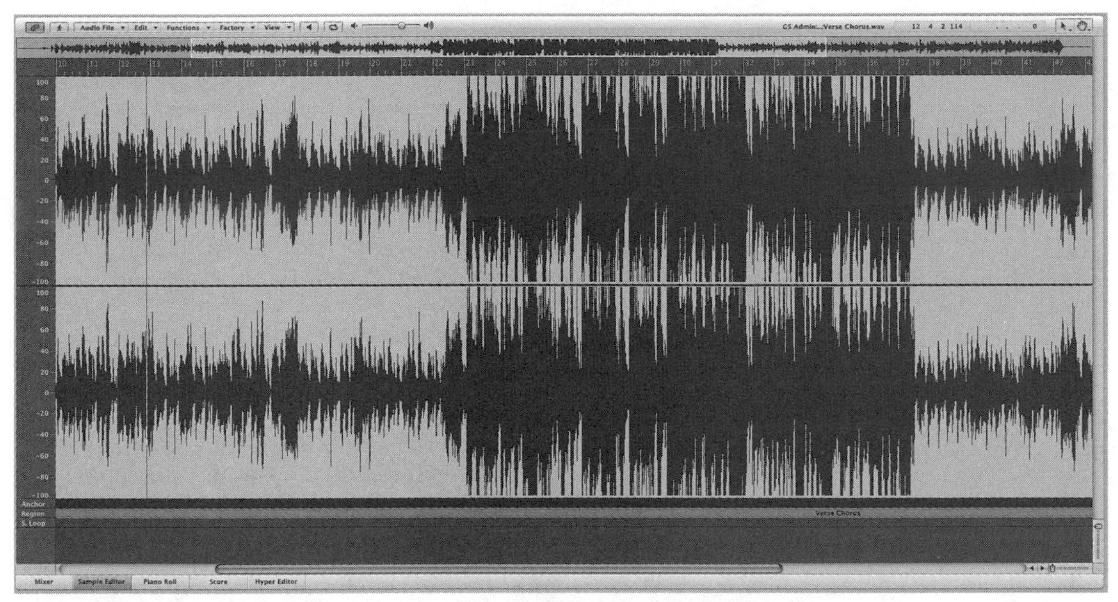

Here's a classic case of poor dynamic control. The verse is almost untouched, while the chorus ends up being heavily limited. Better dynamic control ahead of a limiter would help in this situation.

With this mistake in mind, let's explore a number of ways you can use a compressor ahead of limiting as part of the loudness equation. For a more detailed explanation on how to configure each of these compression techniques, it's well worth referring back to Chapter 4.

Moderating the overall song dynamic

To negate the problems that we've described previously, it helps if compression can start to bring some refinement and shape to the music, ideally lifting some of the detail and body in the quieter parts of the track and then conservatively reducing the levels (as transparently and musically as possible) when you reach the louder parts of the track. Overall, you might not notice more than a few decibels of loudness improvement, but you'll aid the limiter in respect to it needing to do a little less work.

There's a range of techniques at your disposal, including the 'gentle mastering compression', 'over-easy', 'glue' and 'classical parallel compression' techniques we looked at in Chapter 4. On the whole though, you'll want to use the low threshold/low ratio model so that the compressor is working in marked contrast to the brick-wall limiter – massaging the body of the track, rather than attacking the transients.

Loudness: 'Sound Checking'

One interesting development in the loudness wars is the increasing prominence of automatic 'loudness correction' tools such as iTunes' Sound Check. In theory, the algorithms behind systems like Sound Check analyze the loudness of a given audio file and then permanently store this information as part of the file's metadata. On playback, iTunes then applies volume compensation based on the loudness metadata – potentially raising the level of a CD mastered in the 1980s that has a degree of headroom present, while attenuating a modern-day master with aggressive limiting.

With similar 'sound check' technology being adopted in radio (known as *replay gain*), the days of aggressive brick-wall limiting as a means of extracting the last few decibels of loudness may well be numbered. As a result, many engineers and musicians might start to take a step back and think about what makes a recording 'better' rather than 'louder', potentially encouraging technology such as Sound Check and replay gain to turn up their masters! Already there are moves within the audio industry to counter excessive loudness and return to a more balanced viewpoint, including the Dynamic Range Day movement, and an organization called Turn Me Up. Either way, we may soon look at aggressive limiting in the same light as the overuse of digital reverb in the 1980s!

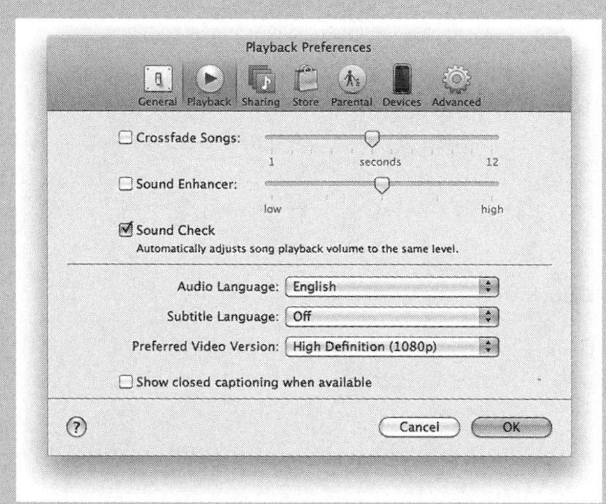

Auto loudness correction technology – like iTunes' Sound Check – is slowly changing the way we look at loudness and the benefits it brings.

Moderating transients with analogue peak limiting

Try to control excessive peak energy using an analogue peak limiter (using the 'peak slicing' technique described in Chapter 4) rather than using your brick-wall limiter. An analogue peak limiter lacks the surgical efficiency of the brick-wall limiter (which is essential to run the mix right up to 0dBFS without clipping), but it sounds significantly more musical. If the track is already well balanced in respect to its transients, this is probably a superfluous exercise.

Small amounts of analogue-style peak limiting can help control transients in a musical way. Analogue limiting won't take you right up to 0dBFS, but it helps control transient energy ahead of digital brickwall limiting.

Controlling the frequency balance with multiband compression

Controlling the frequency balance through multiband compression is fundamentally important given the points that we've noted about frequency and loudness. In particular, some solidity in the bass end will help in respect to both the use and allocation of sound energy, but will also help to create an even dynamic in the bass spectrum so that the bass has a consistent contribution to the low-end. Small amounts of high-end 'hype' can also help mimic the effect of music being played loudly, although care should be taken so that this doesn't become too fatiguing.

Given that multiband compression deals with both timbre and dynamics, it's a powerful ally when working in conjunction with brick-wall limiting in creating loudness. Even so, you still need to take an adaptive approach to understanding the needs of the track you're trying to master, and how multiband compression can enhance them. Unlike limiting, which generally delivers a few extra decibels of loudness whatever you pass through it, the application of multiband processing won't automatically make a track louder – it needs to have an objective to work with.

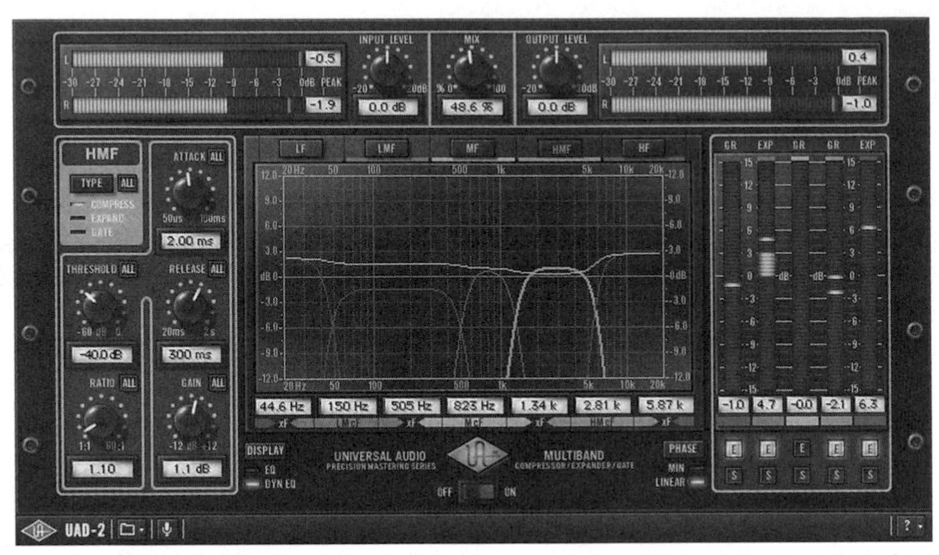

Given that multiband compression works with both dynamics and timbre, it's a powerful loudness tool to use in conjunction with a brick-wall limiter.

6.10 Practical loudness Part 3: Equalizing for loudness

Having looked at some of the broader issues of shaping timbre in Chapter 5, and the impact of frequency on perceptions of loudness, lets look at how we can specifically use equalization to create, or at least better control, loudness.

There are a number of keys techniques to can exploit, each with varying degree of effectiveness given the material you're presented with.

Rolling Off the sub

As we saw in our general chapter on shaping timbre, a 20–40 Hz high-pass filter is de rigueur in most mastering signal paths, so you'll probably already have this in place. However, it's worth remembering that a degree 'sub control' will often help in the creation of loudness, largely as a means of negating excessive and unwanted sound energy usage by the bass spectrum. Of course, a counter argument can be made based on the power of the low-end, as the real depths of the bass spectrum can being used for dramatic creative effect. In these cases, make sure the cutoff is positioned at a suitably low level (even 10 Hz can make a small contribution), but possibly also combine it with a small boost 10 Hz or so above the cutoff point.

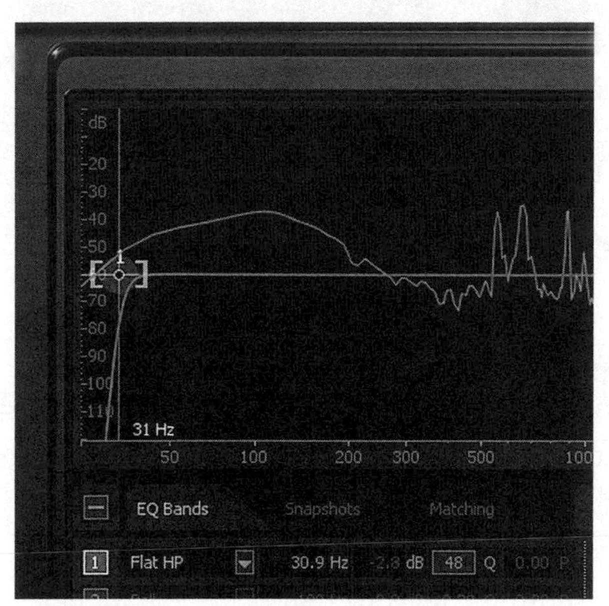

Given the amount of sound energy in the subsonic region it's worth keeping the extreme lows of the mix under control.

The 'smiling' EQ

Formed from a shelving boost at the two extremes of the frequency spectrum and a broad mid-range cut, the classic 'smiling' EQ curve is a well-known sweetening enhancement for loudness creation, mimicking the way in which our ears flatten the frequency response at louder amplitudes. Certainly, if you want to create the effect of music with energy and excitement, this 'smiling' EQ should be considered as an essential move. Be wary, though, of making the music too harsh and adding too much wooliness and weight to the low-end, as neither of these qualities will improve on your master's eventual loudness.

If you're not quite getting the highs and lows from your original material, consider (with care) using some of the excitement and enhancement techniques described in Chapter 5.

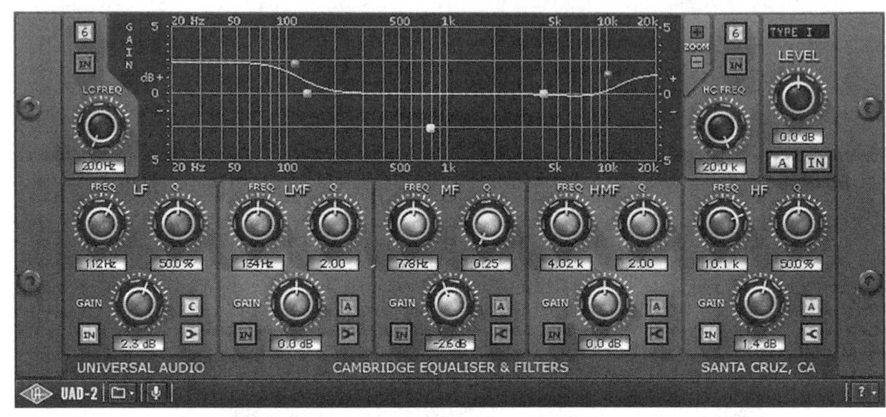

The classic 'smiling' EQ has long been used as means of simulating the enhanced treble and bass perception delivered by 'loud' music.

Controlling excessive mids

Arguably the opposite of the 'smiling EQ' trick, it's worth considering whether any excessive mid-range information is reducing the loudness potential of your master. Remember that the ear is particular sensitive around 1 kHz, and that sharp peaks around this part of the frequency spectrum may inhibit the perceived balance of treble and bass. In short, the track will probably sound 'harsh loud' rather than 'big loud', which isn't a good thing. In the case of an excessive mid range, you're much better applying a broad parametric dip before you go anywhere near treble and bass boosts.

Harmonic balancing

One of the more interesting and scientific applications of equalization for loudness is in the process of creating a more even and unified frequency plot for your master, as exemplified by applications such as Har-Bal. In theory, the technique takes a revised look at signal levels in respect to achievable loudness. So, rather than seeing amplitude as the sum of all frequencies, we see how each frequency contributes to the overall loudness of a piece of music. In short, *harmonic balancing* uses the fact that not all frequencies contribute equally, and that overall peak levels are often dictated by a few stray harmonics being excessively strong.

The first step in harmonic balancing is to gather a detailed frequency analysis plot of your given audio material. The frequency plot can be achieved using a variety of off-line audio editors, such as Steinberg's

WaveLab, various real-time audio analyzers, including Waves' PAZ, and the audio analysis feature of Har-Bal. Ideally, what you need is some form of averaged display, allowing you to identify principle frequency-based peaks, as well as any potential troughs or a mis-balancing of the frequency spectrum. As a guide, it might also be worth referencing a commercial track, potentially noting a generalized pattern towards a uniform frequency response and a lack of distinct frequency-based peaks.

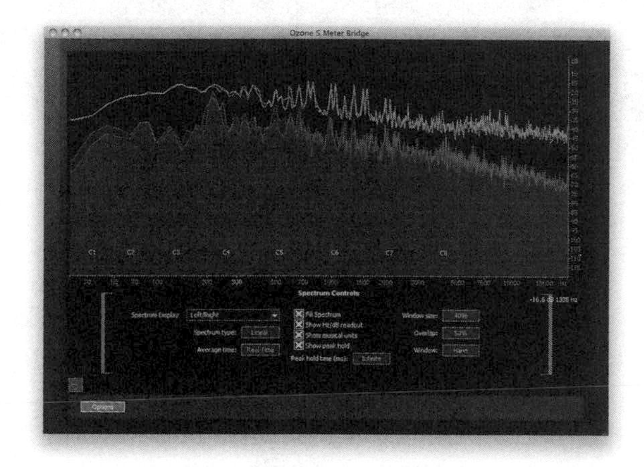

Using a form of 'averaged' spectrum plot, try to find any distinct frequency peaks that occur throughout the full duration of your master.

Once you've identified the 'peaking' frequency points, you then need to start applying equalization as a means of smoothing out these peaks and troughs. Given that transparency is key here, consider using a phase-linear equalizer with plenty of precision, or, if you are using Har-Bal, its own in-built phase-linear equalizers. You'll need to use relatively tight Q settings so that you only modify a small band of frequencies. In theory, by notching these unwanted peaks and raising potential 'holes' in the audio spectrum you'll gain a neutral sound, as well as one that will have the potential to have is overall signal level raised.

While the technicalities of harmonic balancing have some theoretical merit, there's some argument as to whether this is simply an inherent part of a mastering engineer's skill-set and a process that is intuitively applied to create a balanced timbral profile. Certainly, graphical tools such as spectrum analyzers can help identify potential frequency issues with a degree of resolution and accuracy that

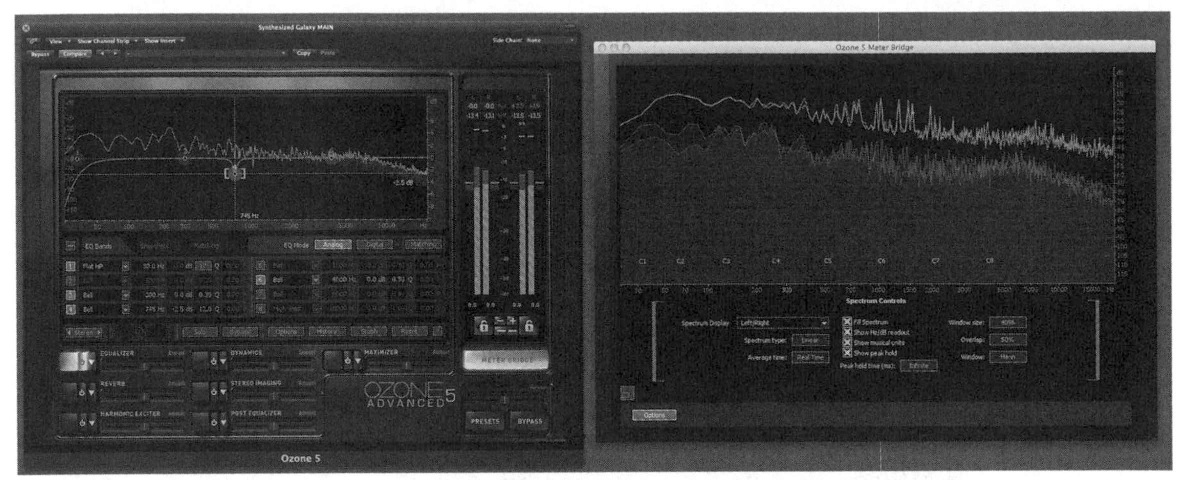

Try applying corrective equalisation to address the peaks and troughs that you've identify in the frequency analysis.

few ears can match, but mastering engineers have worked for many years without such precise tools to hand. Ultimately, a balance needs to be struck between a 'theoretically' perfect sound and one that has the greatest degree of musicality. In that elusive search for loudness, though, the technique of harmonic balancing may well deliver those extra few decibels that make all the difference!

6.11 Practical loudness Part 4: Stereo width and loudness

We've already hinted at the role of stereo width and loudness in Section 6.8, but here's an opportunity to consider how the manipulation of the track's width in mastering can affect loudness. Remember, most of the power of this mix comes from its centre, while the outsides provide colour and texture for the track to sit in. The obvious way to control the balance between these two entities is M/S processing – either through adjustments to the stereo width controls as part of an M/S matrixing plug-in, or by influencing the representation of the M/S channels with compression and equalization.

Although we've touched on various M/S processing techniques before, here're two simple techniques that you can apply to help in the process of loudness creation.

Compress the mid channel

This is where the power of your mix lies, so ensure it has a consistent amount of sound energy behind it. As a by-product you'll probably enhance principle instruments, giving the track more of a direct sound, with less detail and embellishment. The effect might be subtle, but it's a clear refocussing of the direction of the music towards power.

Attenuate bass in the side channel

Bass in the side channel is a somewhat wasted entity given that our ears can't discern stereo information beneath 100 Hz very well. As a result, you'll free up a small amount of additional sound energy, as well as producing a tighter more defined bass overall. A number of M/S processors – including the Maselec MTC-6 and Briainworx's bx_digital V2 – feature dedicated controls for this specific task.

6.12 Practical loudness Part 5: The brick-wall limiter

Despite the importance of a multifaceted approach to loudness, there's no denying that a limiter is still the principle tool for adding loudness to a master. On the whole, the limiter is placed as the last device in your signal path, defining the amount of transient reduction, overall loudness and final programme levels for the finished master.

While the precise design changes from model to model, most limiters work with two principle controls – *input* and *output*. At unity gain – with the input at 0 dB and the output at 0 dB – no limiting is applied. Raising the input starts to drive the gain reduction circuitry, assuming that the mix currently peaks somewhere near 0 dBFS (otherwise you'll simply be raising the overall programme level slightly). As you push the input control harder an increasing amount of limiting is applied to the peak signals, reducing some of the transient energy but also increasing the average signal levels, and hence the loudness.

How much limiting can I add?

The amount of limiting you can realistically apply is determined by a number of factors. One fundamental part of the equation is the limiter itself – some poorer 'integral' DAW limiters, for example, distort quickly and sound unpleasantly harsh, while better-quality third-party limiters sound more transparent and smoother on the ear. The source material also has a part to play. Tracks with plenty of energy and a percussive drive can often tolerate generous limiting, while softer material without percussive bite (such as a solo piano and vocal) need a more conservative approach to avoid obvious distortion.

As a general guide, start with about 2 dB of limiting as a conservative starting point – just enough to lift the average levels and control some of the peaks. If the track that you're trying to process is particularly transient in its nature, it might be that you can push the limiting hard, maybe adding another 2–3 dB of input drive. As you'd expect, consistent amounts of heavy gain reduction will start to make the presence of the limiter felt, whereas lighter or more sporadic limiting can largely pass unnoticed. Indeed, assuming that it's only a momentary excursion, it's surprising just how hard you can push a limiter without it sounding too painful!

Pushing your limiter's input will increasingly drive the gain reduction harder, reducing transient energy and increasing the overall loudness.

If you're lucky enough to have access to a number of different limiter plug-ins, it's worth experimenting with them to see which best suits the track you're trying to master. Despite their relative simplicity, all limiters have a slightly different sound, so it's well worth comparing a few models to see how they behave. Some limiters also feature variable release times, moving away from the usual programme-dependent auto setting, as well as adjustable attack characteristics that suit softer, more acoustic applications.

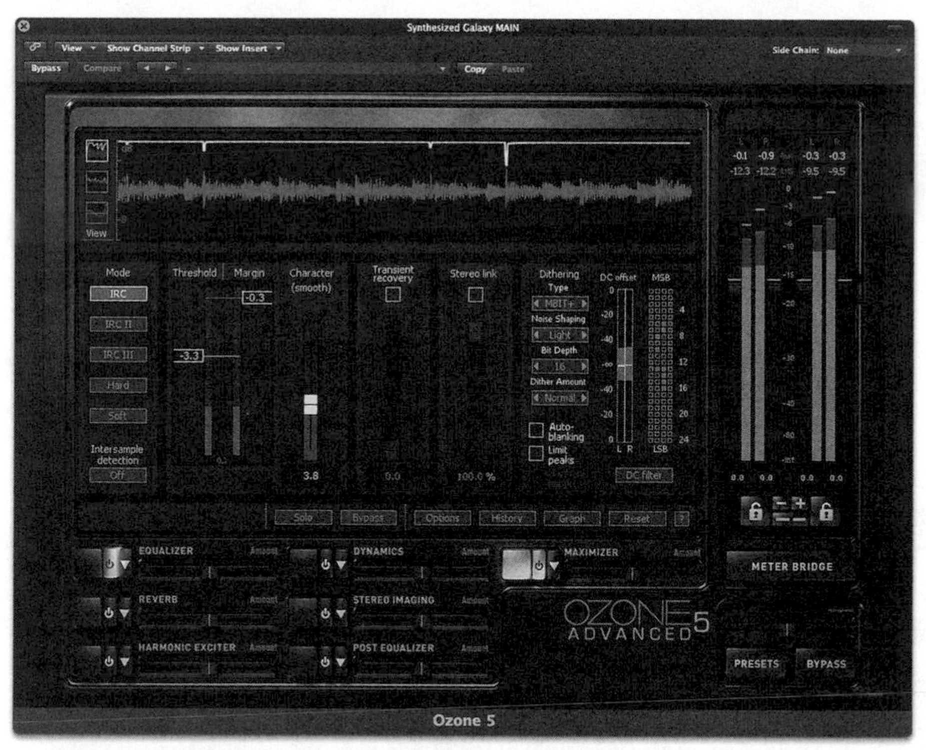

Try comparing different limiters, as each tends to have a slightly different performance and sound.

Knowledgebase

What is a digital over?

A digital 'over' is the oxymoron of the mastering world – an apparent contradiction in terms that seems to make little or no sense. As the theory of digital audio dictates, no signal can exceed 0dBFS in a digital system – put simply, there are no more 'ones' left in the digital stream to encode amplitudes higher than 0dBFS. The contradiction comes in the metering, which often seems to display a digital over, indicating that the signal level has 'exceeded' 0dBFS, sometimes even giving you a figure (+0.5 dB, for example) to work with.

In truth, a digital over is metered not when the signal exceeds 0dBFS, but when the converter receives a number of consecutive full-scale samples. Technically speaking, the logical assumption of consecutive full-scale peaks is that the signal has been clipped, but in reality a square wave normalized to 0dBFS could be triggering a digital over meter reading even though it wasn't technically distorted. Likewise, a heavily clipped recording could be reduced by 0.5 dB and wouldn't even tickle the over meter!

The conclusion? Well, digital overs need to be read with a pinch of salt – just because it says it has clipped it doesn't mean that it's so. If you're trying to ensure signal integrity throughout the entirety of the production chain, it's probably important that you heed the warning that the metering is offering. Ultimately, it's important to use your ears and a sensible splash of headroom in the right places (especially before the limiter) rather than playing consistently up to 0 dBFS.

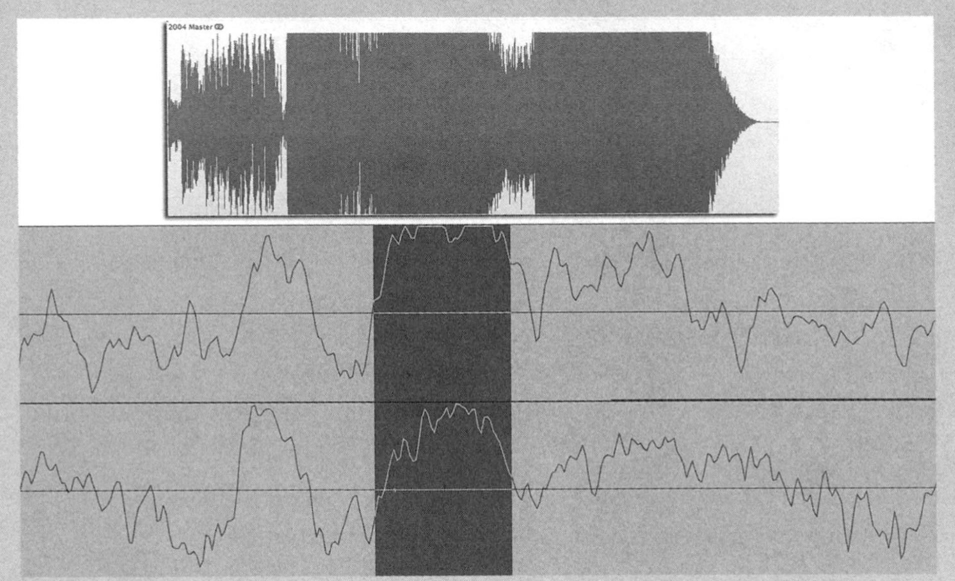

Rather than being a true 'overload' as such, Digital Overs are usually an indication of number of consecutive full-scale peaks, indicating that the waveform has been clipped.

Setting the output levels

The output control sets the final output levels for the master, which, given that the limiter is the last device in your signal path, will form the final signal levels for your production-ready audio file. The output level is important for several reasons. Firstly, despite our apparent need to run a master right up to 0 dBFS, there's a potential risk of so-called 'inter-sample peaks' as part of the D/A conversion process (for more information on inter-sample peaks see the Knowledgebase). Inter-sample peaks can be easily negated by running the output at around–0.3dBFS – a figure that is becoming the standard for peak levels in CD mastering.

In most cases, the limiter will set the final programme levels, with -0.3dBFS the current standard for audio CDs.

Of course, there's also an argument that you might want to run the output lower to create a more balanced overall loudness across a complete CD, arguably in situations where you want the added 'sound and squash' of a limiter driven hard, but not at the expense of a track sounding proportionately louder than the rest of the CD. It might also be that you need to deliver the master at a level slightly lower than 0 dBFS, which is often the case in film and TV sound.

Multiband limiting

For real 'loudness junkies' there's also the option of multiband limiting, as found in the Waves L3-16. The L3-16 works as a form of auto-summing 16-band multiband processor, with a unique brickwall limiter for each band. Unlike a traditional multiband compressor, there's no control to establish the crossover points between the bands. Instead, the L3-16 is simply designed to be as discrete and transparent as possible, only applying limiting to the bands where the peak transients occur. In theory, the L3-16 negates unnecessary modulation effects, allowing you to achieve greater amounts of limiting with less unwanted distortion.

6.13 The secret tools of loudness – inflation, excitement and distortion

As we've already identified, limiting as a means of creating loudness is a distinct case of diminishing marginal returns, with many limiters having a finite point beyond which any increase in compression only ever produces an increasing amount of distortion rather than loudness. Equally, the loudness that a limiter creates isn't appropriate to all forms of music, and doesn't really address the multitude of properties that really affect our perception of loudness.

To fill this gap, there are a number of products that either directly or indirectly create loudness by means other than peak limiting. As you'd expect, the exact 'mojo' that lies behind them is often never precisely explained, although to our ears, most sound like a form of body-enhancing parallel distortion and compression. The subtle use of distortion makes a lot of sense, given its harmonic reinforcement (particularly on bass), dynamic saturation and a touch of high-end sizzle. Ultimately, it lends a master slightly more weight, which can really help in reducing how hard you have to run the limiter at the end of your signal path.

Although not exhaustive, here's our list of recommended plug-ins, along with some basic descriptions of their use and operation.

Inter-sample peaks

Inter-sample peaks are a product of the D/A conversion process and are a product of our belief that the final output levels directly correlate to the peak meter readings inside our DAW. As part of the conversion process, reconstruction filters effectively round-off, or interpolate, the stepped digital output into a smooth continuous analogue waveform. As a result of the rounding-off process, it's possible for the final 'analogue' output to be slightly higher than 0 dBFS. For a well-designed converter, this extra level isn't a problem, as headroom will have been factored into the equation. However, for a cheaper converter the excess level can create distortion, otherwise known as inter-sample peaks.

The issue of inter-sample peaks is further compounded by its unpredictability. While the master might sound fine over a mastering-grade converter, the same might not be true of a cheap CD player, or the soundcard of a cheap PC. In short, there's no way of telling exactly how the track will behave in a range of listening environments, and dedicated inter-sample peak meters (such as SSL's X-ISM) only indicate a likelihood, rather than an absolute.

To avoid the potential of inter-sample peaks there are two options. Firstly, try creating half a decibel or so of headroom – just enough to negate potential distortion, but not so much as to unduly affect loudness. Secondly, look for a limiter that incorporates some form of inter-sample peak prevention technology, such as Ozone's Maximizer.

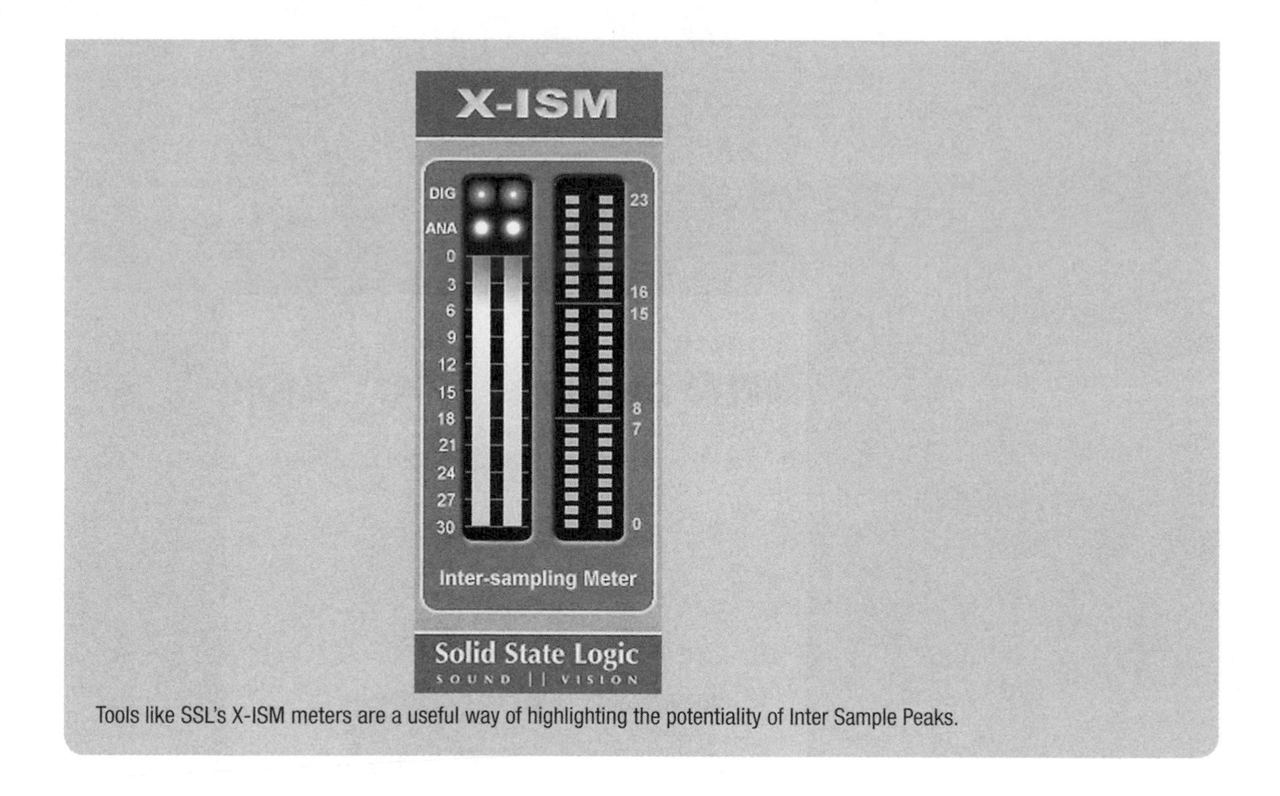

Tools like SSL's X-ISM meters are a useful way of highlighting the potentiality of Inter Sample Peaks.

Sonnox Inflator

Sonnox's Inflator was one of the first plug-ins to be released as a true alternative to limiting as a means of loudness creation. The two principle controls are *effect* and *curve*, although it's also possible to overdrive in the *input* to the plug-in (either to restore gain, or to provide deliberate distortion) as well as being able to define a final output level. The best way of setting the *inflation* effect is to start with the effect parameter at 100 per cent and then experiment with the curve control, which somewhat changes the colour of the inflation effect. To achieve a reasoned amount of inflation, back down the effect percentage so that the plug-in adds the right amount of 'inflation' without the programme material sounding too coloured or distorted.

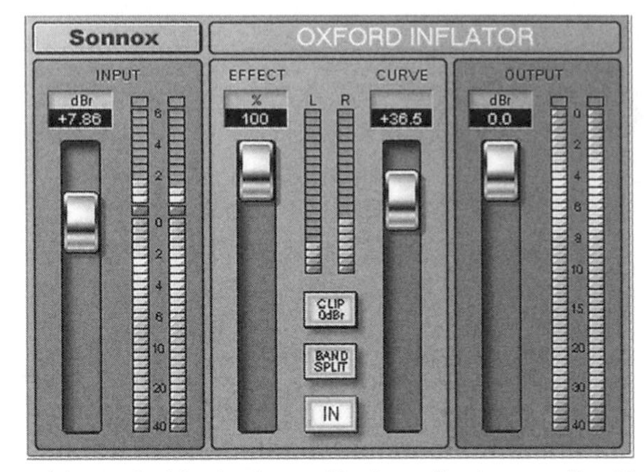

Sonnox's Inflator is useful way of adding loudness without resorting to conventional transient-reduction techniques.

Slate Digital Virtual Tape Machines

Slate Digital's Virtual Tape Machines is one of the best plug-in emulations we've heard for recreating the 'sound' of tape. Unlike some tape emulators, Virtual Tape Machines makes a point of emulating both a 16-track 2-inch recorder, as well as a 0.5-inch mastering deck running at 30 ips. In many ways, tape represents the ultimate form

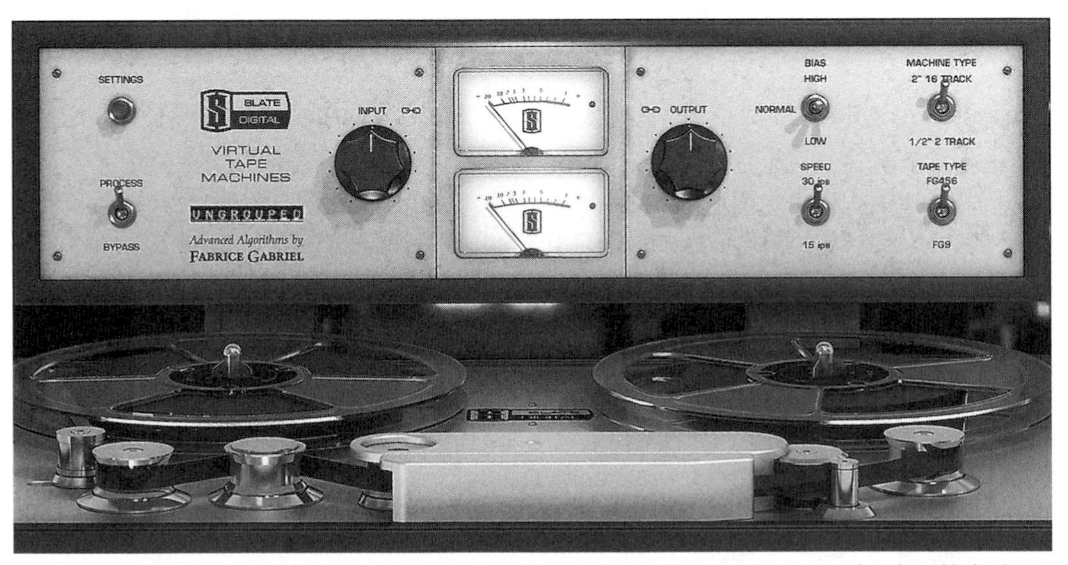

As we've seen, tape saturation can be a useful way of controlling transient energy, but the added colour and body of tape saturation can also make a track feel louder.

of loudness creation, mixing both elements of transient reduction, colouring and saturation. However, it's a technique that varies in its effectiveness given the material your working with, so it's best to try experimenting to see whether it adds or distracts from what you're trying to achieve. As you'd expect, acoustic music often benefits from the 'softness' of tape, but hard-edged digital production styles often loose some of their precision and edge.

Universal Audio Precision Maximizer

Universal Audio's Precision Maximizer arguably forms a halfway-house between the classic effect of tape saturation and the 'inflation' effect described above. In effect, the plug-in is pitched as a form of 'audiophile distortion', modelling the pleasing body-enhancing non-linearities of an analogue path. As a result, the Precision Maximizer creates both a proportionately louder output, and adds some pleasing tonal qualities in respect to extra bass body and a touch of sizzle. Like the Inflator, the main two controls cover *shape*, which is somewhat equivalent to the amount of drive, as well as *mix*, which sets the amount of 'Maximized' signal blended into the final mix.

The Precision Maximizer shares many similarities with the use and operation of Sonnox's Inflator, and can be a great way of enhancing loudness in a more 'rounded' way.

CHAPTER 7

Controlling Width and Depth

In this chapter

7.1 Introduction 169
7.2 Width 170
7.3 Surround or no surround? That is the question........ 171
Surround-Sound formats and current mastering practices........ 171
7.4 Stereo width........ 172
7.5 Phase........ 172
7.6 M/S manipulation for width 175
7.7 Other techniques for width 176
7.8 Depth 178
7.9 Mastering reverb........ 179

7.1 Introduction

In previous chapters we have discussed aspects of dynamics, timbre and loudness. However, we've overlooked one area that probably deserves most of the discussion: space. Spatial factors add to the overall excitement of a music production and are something the mastering engineer can manipulate to enhance, or destroy!

We've broken these spatial elements down into two broad descriptors: width and depth. We don't discuss height in terms of up or down, as this is rather difficult to achieve sonically with two speakers. Rather, for the purposes of this book, we refer to height in terms of high to low frequencies.

Given the descriptor 'width', and also following the ethos of this book, this chapter concentrates on stereo and the spatial manipulation between two speakers, rather than the more literal surround-sound. Surround-sound mastering will be an extension of the knowledge here, and contains some slight differences that we'll touch upon from time to time, but not in extensive detail.

It never ceases to amaze how two speakers can envelop a listener and provide a wholly developed sonic landscape. With the current trend towards high-quality headphones, never has it been more important to understand stereo and how the work we do can affect it. With depth we explore the ways in which engineers can create three dimensions to a mix at the mastering stage.

7.2 Width

As with any new format or practice, there is always a little bit of time needed before conventions settle down and become the norm. It is fascinating to see this in the history of stereophonic sound. Listening back to many of the earlier Beatles albums, knowing that stereo was something of an afterthought to the Abbey Road engineers explains why the mixes separated out the drums and vocals across the two speakers.

Understanding how they tracked the material on a four-track tape machine goes someway to explain this. They would be left with four individual tracks to mix that had been recorded with no concept of stereo as a two-track phenomenon. To use the format properly, all that could be done would be to separate out the parts presented in the four tracks. Technology such as pan controls and stereo effects units had not been fully explored. If you were to mix material this way now, you'd find it a little strange to listen to, yet those early stereo mixes of the Beatles are simply historic now, and despite the mono ones being the tried and tested (often more popular to the fans) ones, the stereo mixes endure.

Over the period of the band's career, it is astounding to think that within ten years, audio engineering had developed by such a considerable amount (from four track to 16 track, for example) to go from mono to widespread stereo. Abbey Road, the Beatles' last album recording, is awash with the now conventional use of stereo, and is done very well in our opinion. The convention was, to one extent or another, sealed by that point, and in a very short space of time.

7.3 Surround or no surround? That is the question

Surround sound has had a tricky time of it, in music production anyway. Around 15 years ago, there was a widespread ambition to move away from the confines of CD to higher-resolution audio, and to also spread away from stereo to 5.1 (certainly as a format for home listening). However, for important reasons, both technical and social, this has passed away from mainstream focus… for now.

The same thing happened in the 1970s with 'quadrophonic'. Quad, as it was referred to, was equally exciting, with the format being developed to be read on special record players. By rights the four-speaker format could have really taken off, but for some reason it did not really penetrate the mass market, just as 5.1 surround-sound (for music anyway) failed to do two decades later.

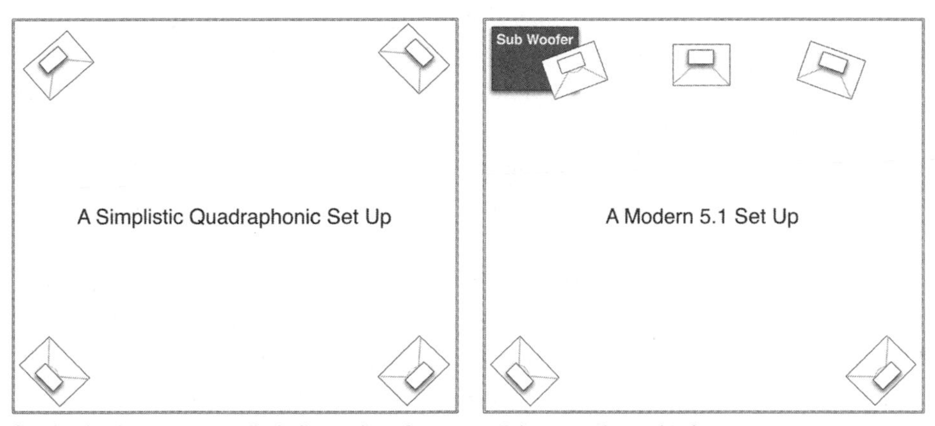

Quadraphonic was not too dissimilar to that of common 5.1 surround sound today.

It is surprising that modern surround sound carried by formats such as SACD and DVD-A never took off. There is been no doubt as to the success of surround sound within cinema, and if you look at the history, it has been installed for some considerable time – since the 1940s in one form or another. Before moving to discussing spatial parameters in stereo, let's look quickly as differing conventions for mastering in surround sound.

Surround-Sound formats and current mastering practices

While current practices demand ever louder stereo masters, the story can be quite different within surround sound. All the other concepts remain largely the same, but you've now got five full-range speakers to spread out the same audio material that was once squeezed into two speakers. Vocals can dominate one monitor (usually the centre monitor) if you so choose. Many differences are probably more in the mixing conventions, as we saw with the introduction of stereo all those years ago.

You'll still wish to apply compression or limiting to your masters in surround sound, but you'll probably feel less pressure to make them sound as loud as the next track, as the dimensions are different. You've got space, and it is likely that the mix will come to you in a less compressed fashion. You'll be able to retain dynamic range and power, plus the mix engineer will have had less trouble mixing as there's simply more room.

7.4 Stereo width

However, within stereo the loudness habit still endures, and we have a lot of music to squeeze through the left and right monitors. For music delivery, stereo still rules. This is even more likely to persist given the ever-increasing popularity of headphones and portable audio players such as the iPod. With this in mind, let's concentrate on stereo mastering and what we can do to manage and enhance this aspect of audio.

There is no secret in the fact that some of the most powerful mixes are in mono – straight to the point, direct and driving. However, the ear also loves the enhancement stereo provides, when it is done well. Mixes that come to us vary tremendously, from those with carefully planned and managed stereo fields, through to those that innocently send us mono bounces by accident (or for real!). Knowing what is required for the track in question is the first thing, but it's also important to understand what stereo is and what it is not.

7.5 Phase

Stereo sound is, by rights, anything that could come out of two speakers, but that's hardly stereophonic sound. Stereo is a

Stereo metering

Metering in mastering is very important, especially for monitoring levels (and now loudness). However, the information about phase and stereo width are also absolutely key to your mastering and ought to be referred to frequently.

The first and foremost meter in this task would be the simple correlation meter. The correlation meter should be employed to keep an eye out for the phase coherence of your audio material. Ideally, the meter should remain in the positive scale, but may wander from time to time into the negative due to deep and expressive effects (such as a phaser over the whole material). Keeping an eye on this can be very useful. Perhaps a piano sample used for a track might have a wealth of DSP on it to make it feel full and warm, but the phase coherence from this might be skewing the correlation meter to the negative.

Most useful to a mastering engineer is the goniometer, or vectorscope, which dynamically shows the phase relationship between the left and right channel. It also provides an intuitive view of the audio which morphs to the music. The information it provides is more of an impression of stereo against phase.

The Correlation Meter is the standard phase coherence meter which you should keep on the desktop whenever you can, although you'll instantly hear phase issues with experience.

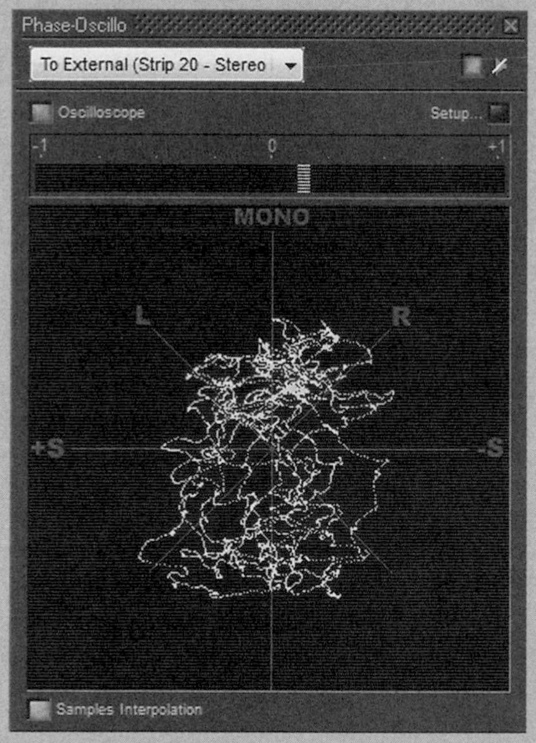

The Goniometer or Vectorscope is key to seeing dynamically how phase and stereo width are working together.

fascinating and enduring medium whereby a phantom mono signal can be created by two equal signals on either side (equal to both left- and right-hand speakers). The result is that you believe a vocal comes from the middle. Stereo is best when the manipulation permits an extremely wide soundstage through the clever use of panning, and in some cases phase manipulation.

Phase coherence is important to mastering, as it provides information as to whether the programme material you're preparing will work on the widest possible audio playback systems available. There's always the argument that your stereo master should naturally fold back to mono with no real detriment. That is one reason why professional monitor controllers and mixing consoles are adorned with a mono button. The other reason is that it is a quick way to check the phase of your stereo material. If that bass drum happens to reduce drastically in size when you put the mix into mono, then you know automatically that there's a phase problem, if you'd not heard it beforehand.

Many believe that it remains important to respect the maxim that a mix should transfer to mono, for some very valid reasons. FM radio, while it is still with us, can sometimes pop into mono when the signal is not strong enough, and would suffer accordingly. Secondly, there are still many people listening to radios in their kitchens, and televisions too for that matter, that remain mono with only one speaker. As such, if your mastered material is to be played out, then you should consider ensuring mono-compatibility.

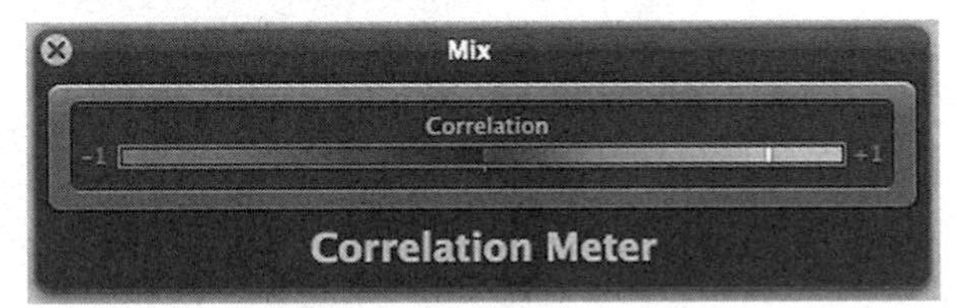

A Correlation meter permits you to visually see the phase coherence of the stereo material you're mastering.

There are many wide mixes that will have your correlation meters jumping towards the negative portion of the scale. This can be due to poorly mixed aspects or due to intended intense production techniques that have created something sonically immersive. Either way, you will need to make a call on how to address this in your mastering, if at all. Techniques for managing some of these aspects are covered below, and are interpreted in the next chapter.

7.6 M/S manipulation for width

You'll have noticed by now that M/S (mid/side) processing crops up throughout this book. It has to. M/S is an extremely valuable, almost hidden, tool in the mastering toolkit, as it allows us to dissect a mix using different dimensions to that of left and right, high and low, quiet and soft. M/S gives us the power to separate the mono from the stereo audio of the recording in front of us.

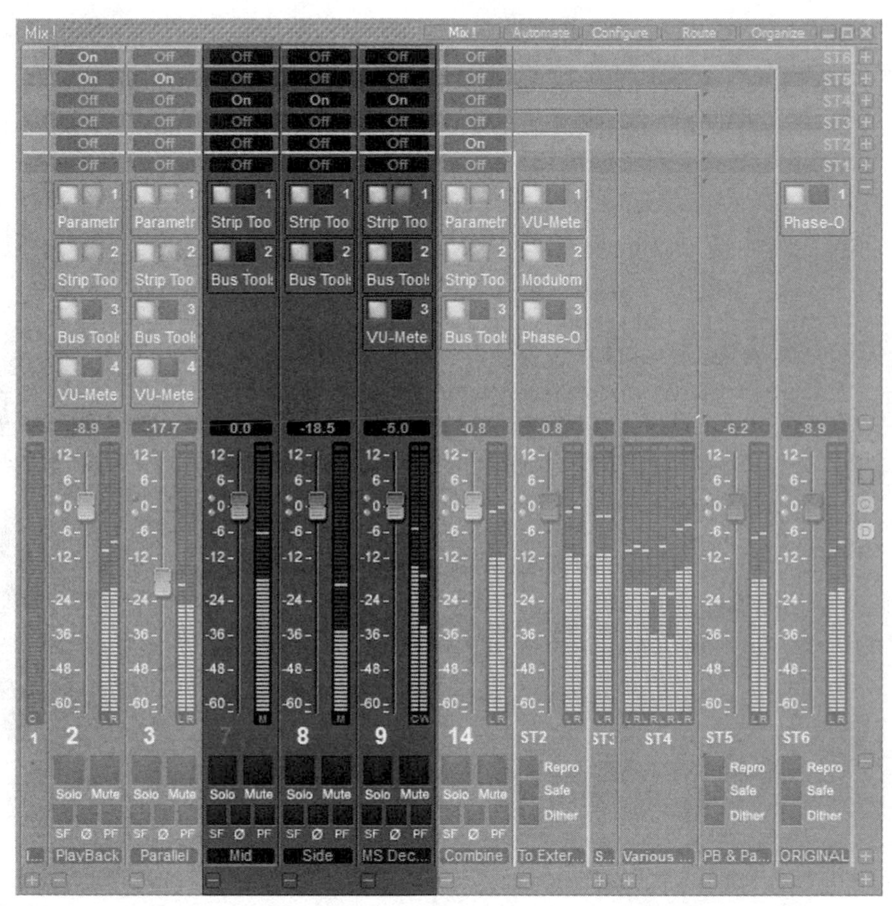

Turning down the mid level here in Pyramix will bring out the left and right information making the sound much wider, but might in turn lose the track some power from the mono (mid) information.

Should we feel that the stereo width is not evident enough, we can simply turn up the side information to reveal more of the stereo signal, or conversely turn down the mid material. Similarly, using M/S we're permitted the possibility of employing EQ or compression on either the sides or the mid to achieve a desired outcome, as has been explored in previous chapters.

Employing M/S manipulation can be incredibly powerful in enhancing the overall power (by processing the mid information independently, perhaps through compression) and spatial aspects (by high-pass filtering the sides and enhancing the 'air') of the mix. However, this technique can be quite addictive and should be used sparingly and as the track requires.

7.7 Other techniques for width

Within many DAWs it is possible to artificially adapt the stereo field, or indeed the perception of it. This can be achieved through the employment of the appropriate widening plug-in. Both Logic and Waveburner offer the same stereo-spread tool. This is a plugin which separates frequencies into the left or right channel.

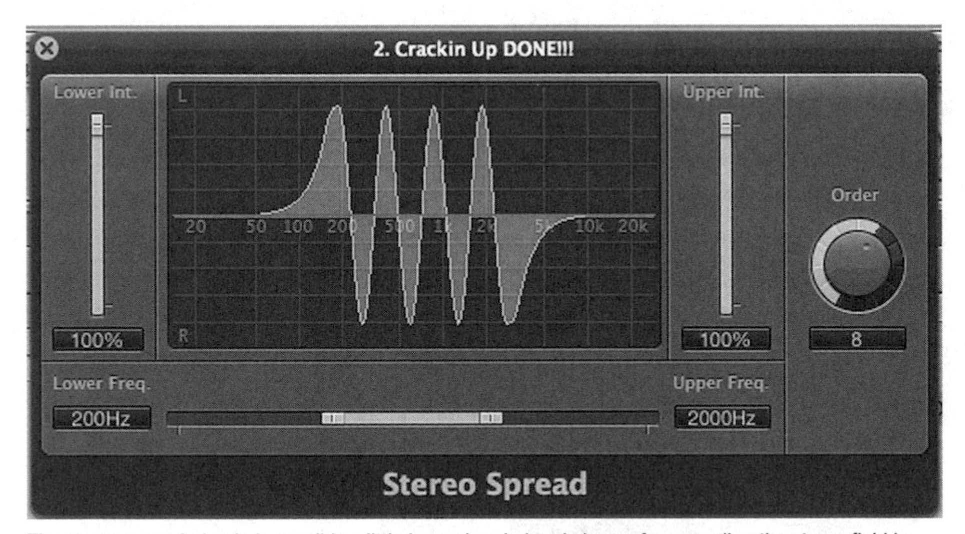

The stereo spread plug in is possibly a little heavy handed and clumsy for spreading the stereo field in many masters.

Other techniques exist to deal with the stereo width of some material. One classic example is to use stereo delay with very short delay times. The result will be some ambience around the image, and if the details are adjusted appropriately some stereo width can be created. Using these small delay times allows us to make use of the Haas effect. Delays of up to 30 ms can be used to increase the concept of space without the ear considering this to be a discreet reflection, although we'd probably advise somewhere below 15 ms.

Software manufacturers offer a number of dedicated plugins to assist with the stereo space issue. One such example is PSP's Stereopack, containing four very useable processors to enhance the stereo field. The first is the Stereo Analyser, which is very similar to the meters found in many DAWs. The plug-ins that massage the stereo field are provided in three flavours. The first is Pseudo Stereo, the second is Stereo Controller and the last is Stereo Enhancer. Each provide their own method of creating or enhancing the stereo image.

As the name suggests, Pseudo Stereo is about creating a stereo image from a mono, or near-mono, source. By suggesting a centre frequency, Pseudo Stereo can then use phase relationships around this to make a stereo field. It even has an emphasis slider to enhance this frequency and bring it out in the mix.

Stereo Enhancer is a little like the Pseudo Stereo in terms of controls, but instead of making things stereo which are not, this simply looks

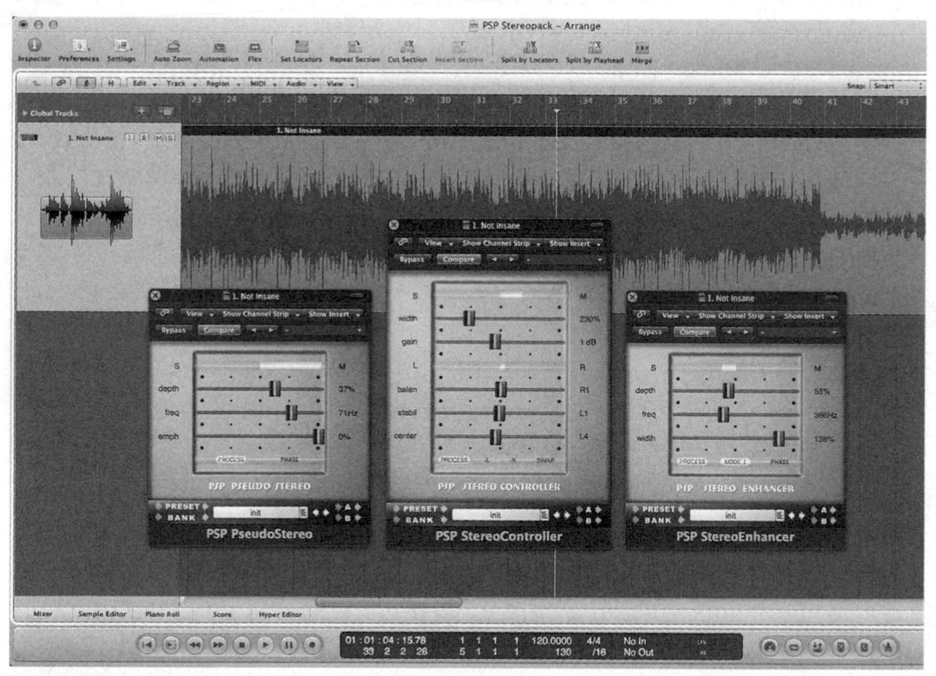

PSP offer four neat stereo solutions in their Stereopack which could be employed to enhance your stereo field.

at enhancing the stereo image that is there so that it fits with your expectations. Finally, Stereo Controller looks at the management of the stereo field and has come quick and simple sliders to enhance or shift the stereo image.

7.8 Depth

Depth is more conceptual as there's no dedicated slider, or direct widget, that can move a particular sound or frequency back and forth in the mix – certainly not for mastering anyway. To a large extent this 'parameter' will be pre-determined by the mix engineer and what you receive, but nevertheless it is something to pay attention to as there are some manipulations you can apply.

Whether an instrument comes from the front or back of a piece of music is pretty subjective in the most part, as we play on the effects that occur when a sound is distant – it will tend to lack definition, have a higher frequency content and be quieter with more reverberation. These are all tricks that we can play in the mix.

In mastering it is more difficult to achieve any form of depth in this way with particular instruments, unless individual stems (for example, the vocals in a separate stereo file – see Chapter 8) are provided, as you'll affect something else in the mix. Similarly, it can be more difficult to retain a delicate balance of instruments creating this depth if you're then required to make a very loud master. The delicate balance between the reverb and the main material could be altered, thus jeopardizing the overall sound.

In the passing of history, there have been many effects that have assisted the three-dimensional presentation of recordings. Two instantly spring to mind: Q Sound and Roland Sound Space (RSS). However, both are effectively unsupported today, and do not see the light of day as they did in the early 1990s on such albums as *Amused to Death* by Roger Waters and *The Immaculate Collection* by Madonna.

Both Q Sound and RSS are algorithms that replicate the way we hear using phase relationships and directionality. For example, a sound from the left-hand side will arrive at your left ear before the right-hand one, thus the phase relationship between the two channels would be very slightly different as a noise is delayed in one ear. Additionally, the sound will have had to get around your head to the right-hand ear, and as such this will have a bearing on the amount of high-frequency content.

An additional factor to take into account is the ear's pinna – the malleable protruding skin from around the ear. The pinna has a number of features. Firstly, it 'funnels' sound coming from in front of you into your ear canal. Secondly, and more importantly for this chapter, the pinna assists the ear in understanding directionality. Sounds coming from behind you will be subjected to phase displacement if arriving from the left- or right-hand side, but will also be subjected to a loss of high frequencies, thus, it is believed, telling you that the sound is coming from behind you as opposed to in front of you. Finally, the pinna assists the ear in understanding whether a sound is coming from above or below due to the pinna's shape and the way sound is funneled into the ear canal.

7.9 Mastering reverb

Adding depth can be achieved by employing a reverb unit. However, a generic effects unit, or plug-in, might not be the best way forward – mastering reverb can be employed to assist in the depth challenge.

On balance, mastering reverb is not employed too often, but is a tool that can be wheeled out on occasion. The problem with reverb on a master is that any broadband effects unit will add reverb to the

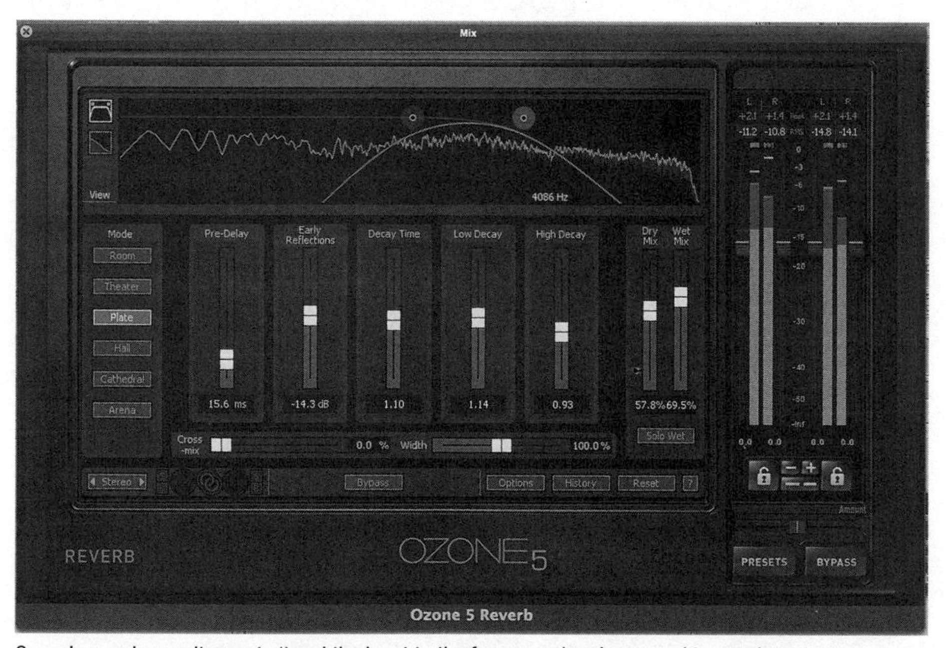

Ozone's reverb permits you to 'tune' the input to the frequency band you need to reverberate.

whole mix, bass drum included. Naturally this is not desirable unless you're simply adding some minor ambience to an extremely dry mix.

iZotope Ozone's Mastering Reverb unit may be lacking in a wide selection of algorithms, but does have some very useful mastering features, namely the ability to choose the band of frequencies you wish to add reverb to. Using a couple sliders you can audition the pass-band (the frequency band the reverb will hear and process) and adjust this according to the area you wish it to add reverb to. The same effect is possible using a standard DAW auxiliary, some filters and a normal reverb plug-in.

More often than not this is applied to the vocal portion of the audio. Sometimes a little 'ambience' is all that is needed, and to many the change may be imperceptible, but this ambience will lift the vocals and make them sit with more presence in the mix, or settle down and bed in.

Reverb is often employed to help out towards the end of a track in the fade-out portion, where a little ambience is needed. This can help the track fade out in the most subtle of ways and blend properly into the next track (if that is what is required).

Other applications might include the mastering of live material where much of the recording has come as a mix from the front-of-house desk and perhaps with a couple of mics picking up the ambience and crowd reaction. Not the best quality, but reverberation can assist here to help bring some cohesion to the diverse sounds from the desk and the audience.

Employ reverb when you need it, but do be aware of the encapsulation it will have over the broadband material affecting bass drums as well as other more reverb-friendly aspects. Take care not to swamp a mix, and also be careful to place the reverberation at a place in your processing chain where it will achieve the required effect. Placing a reverb before the compression for example, will ensure that it 'feels' part of the mix rather than a bolt on, but any dynamics processing later could make this louder. Placing a reverb after compression can provide a sense of dynamics whilst retaining the loudness. It is best to audition it in various positions in the plug-in chain as required.

CHAPTER 8

Crafting a Product

In this chapter

8.1 Introduction 181
8.2 Sequencing 182
8.3 Topping and tailing 183
8.4 Gaps between tracks and 'sonic memory' 184
8.5 Fades on tracks 187
8.6 Types of fades 189
8.7 Segues 190
8.8 Level automation 191
Knowledgebase 192
8.9 Other automation and snapshots 193
8.10 Stem mastering 194
8.11 Markers, track IDs and finishing the audio product 196
8.12 Hidden tracks 199

8.1 Introduction

To many, the art of mastering is often appreciated for the processing it applies and the effect this has on the albums we produce. However, a mastering engineer handles much more than the sophisticated aspects of signal processing. While we acknowledge that this is a critical element, there are less glamorous aspects that require our utmost attention in mastering.

The first of these is editing. To help bring together the larger body of work, whether that is an album or EP, it will be necessary to engage in some rudimentary editing. Tracks may need their starts and ends adjusting to omit extraneous noise at either end of the audio.

However, there may be times when tracks will require a little more of you, from forensic-style editing and removing the odd click, through to segueing one track into another.

In this section we take you through the process of editing your album together: from fiddly editing necessities, the all-important gaps between tracks, and even some noise elimination tools. Finally, we'll complete the section with an increasingly popular way of approaching mastering called Stem Mastering.

8.2 Sequencing

As we discussed in Chapter 3, it can be good to know the order of the songs as the client wants them. If you know this before you begin to process the album, it can inform some of those processing decisions, but if no order has been decided then the processing of an album can have a bearing on how tracks sit next to each other.

Sequencing an album can be harder than you think. It is good to start by getting a feel for how the tracks fit together without any processing. If you've been given the track order, then you can perhaps see where the album flows and where it does not. Making this sequence work will require many of the skills described below. It is a blend of silence, duration and tone. Managing all of these elements in your edits will make an album flow well.

Alternatively, you may be fortunate enough to be asked for your advice and guidance on how the album should be ordered. How to handle the choices and subsequent edits are often decided through experience, and something as simple as an educated gut feeling regarding how something should sound.

The sequence is much more than just what song should sonically follow the one before. Perhaps there's a story to tell? Perhaps there's a sentiment the producer or artist wants? Or perhaps there's a song that, to you, feels as though it ought to be placed at the very end of the album as a sort of 'signing off'.

The tools in your armoury, to make any sequence work, are the raw materials of mastering – level and tonality (EQ) – the gaps between tracks, plus the fades on intros and outros. These aspects and their manipulation, discussed in this chapter, can have a critical bearing on how you put an album together, and also on your ability to experiment.

By cueing two tracks together and shutting your eyes as you listen to a minute or so of the end of one track to its fade out, note the point that you think the next track ought to start. This can be a good 'hunch' as it will be suggesting a number of things to you (the tempo of the last track, the type of fade and your anticipation for the next track) that will have a bearing on where you, the listener, would ideally expect the track to begin.

It is worth using trial and error. Try new gaps, fades and ideas to see what the limits are to this process – what works and what does not? Over time this will build into a personal 'library' of the gaps and edits which you might wish to explore with future masters.

8.3 Topping and tailing

The first things to concentrate on are the most rudimentary of editing tasks. Very often the material will simply need some topping and tailing – the terms we use to chop excess silence away from the start and end of a piece of music.

Many engineers, rightly, will not concern themselves with tight edits at the start or end. In many instances, mixes coming to you will have very large fadeouts which can be re-faded and made shorter in the mastering studio. Repeat-chorus-to-fade songs might just allow choruses to repeat in the mix until the musicians stop playing again, leaving the mastering room to do the 'repeat to fade'.

It is worth taking care here. All too often it is easy to make snap decisions based on the waveform drawn by your DAW. Be sure to listen and 'scrub' the starts and end to make sure that you've not inadvertently chopped off some quiet introductory sounds at the start of a track. It is also desirable, if the audio permits it, to leave a little air before the audio starts. By this we mean the silence in the recording rather than the absolute silence provided by a blank track on a DAW. Something around a second or two would suffice. For those tracks without air, it can be useful to extract a portion of silence, which is actually air from the studio, and place at the top and tail (start and end) of a track without it, to give the impression of cohesion across an album.

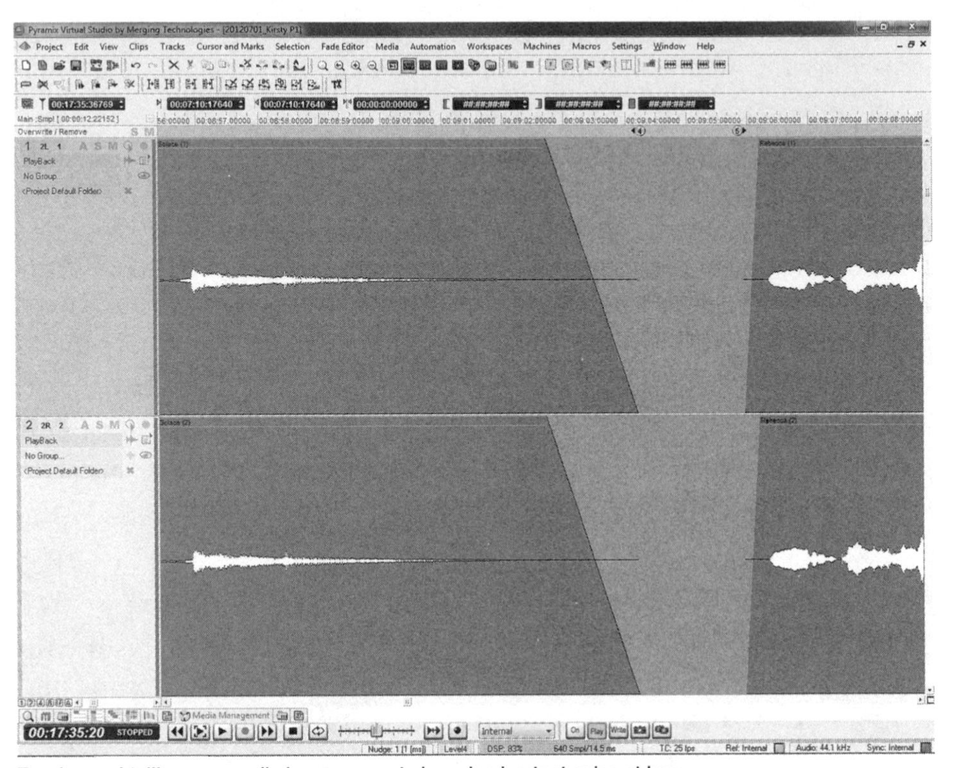

Topping and tailing your audio is necessary to keep inadvertent noise at bay.

8.4 Gaps between tracks and 'sonic memory'

After topping and tailing the audio tracks that make up your album, time ought to be devoted to the way in which these tracks relate to each other. For some reason, a nominal two-second gap seems to be accepted wisdom between two tracks on an album these days. By all means accept this a default, but as with everything, rules are there to be broken and we've rarely found two seconds to be a best fit!

Gaps between songs can be difficult to specify arbitrarily. Even using our analogue and 'rough' method of clicking fingers or tapping the table (our choice when we feel a track ought to start!), gaps can be perceived, if not determined, by the fade out of a previous song. Take a fairly melancholy song, which has an eternally long fade-out of anything up to 30 or 40 seconds. After such a song, where the audio essentially peters out, it's quite plausible to have a new loud track start relatively quickly afterwards.

One concept that could inform your decisions when spacing tracks is what we call the 'resonance' that your ears hold from the last piece of music heard. The word resonance could be replaced with 'tone' or even 'feeling', but here we refer to it as *sonic memory*. For example, let's say you have a loud metal track that ends quite abruptly with a crash cymbal and fades out fairly quickly. Your brain sort of retains a 'tone' or 'feeling' or memory of the piece.

There's almost an interpretation that the 'sound' continues. Now, with the loud metal track one can either start another loud metal track, probably around two seconds afterwards (two seconds being the aforementioned standard gap introduced between tracks by many programs) or, given the 'sonic memory' the previous track provides, the next track could start a little later as one's ear is 'still listening', as it were.

Returning back to our example with the long fade-to-silence track over 30–40 seconds, it is quite plausible that the ear may have given up on the tune by the time it has got to the end. Therefore, you could startle the listener by placing a loud track within milliseconds of the fade out. This could be desirable depending on the effect you wish to create. Being aggressive, it would be ideal if the intention is to startle the listener, but on other occasions it might be best to make the sonic memory die down and settle itself before entering into a new piece.

Alternatively, you could feel your way and, as before, estimate where the track should actually sit. We would tend to advise that you to move away from the visuals that the waveforms can provide on the screen. Your fans and your client's fans will almost certainly not be looking at waveforms unless they happen to be listening in SoundCloud. You ought to do the same. When listening try, whenever you can, to either look away from the monitor or turn it off completely (minimizing the window is a quick way to achieve this).

It would be wrong for us, at this juncture, to give you any hard-and-fast rules for handling gaps. Below is an extremely rough table about how gaps could be approached. By no means should they be considered as dos or don'ts, but something you might like to experiment with in your mastering. After all, are there any right or wrong answers within the world of mastering?

Gap time	Comments	Examples
Immediate/ Segue	Immediate gaps suggest either a segue or something of a shock tactic. A classic example would be a quiet and perhaps long ending of the first track and a loud snare hit abruptly arriving earlier than you'd expect.	The transition between 'The Power of Equality' and 'If You Have To Ask' on the Red Hot Chilli Peppers album *Blood Sugar Sex Magik* is a fine example of how the power of an immediate nil-gap can enhance the flow of an album. Another example is from *6: Commitment* by Seal. The transition between the opening track 'If I'm Any Closer' segues beautifully into the next track 'Weight Of My Mistakes', with a slide down to the correct notes for the second track.
Less than 1 second	While still being almost immediate, there are many tracks that will fit well in this section, from those wishing to cause an impact, to those that sit neatly together due to a storyline.	As far back as 1980, AC/DC used the short gap to keep the pace going on side two of their LP (later transferred to CD) between 'You Shook Me All Night Long' and 'Have A Drink On Me' from their *Back in Black* album.
Between 1 and 2 seconds	Gaps possibly fall into this region most commonly. This can feel most natural as a transition between similar-sounding tracks, or those wishing to keep 'pace' on an album.	Between 'Rolling In The Deep' and 'Rumour Has It' there is a short gap which keeps the pace of Adele's album *21* flowing very well indeed.
2–3 seconds	Two seconds is the 'standard' gap, although can be tailored to suit each track. As you move to three seconds a disconnect begins to occur, meaning that you can place pretty much any tracks together, as it provides the listener with enough down-time.	The songs 'Shiver' and 'Spies' on Coldplay's *Parachutes* have a 2–3 second gap and this feels perfectly natural and normal between tracks such as these. A default perhaps?
3 seconds or more	More than 3 seconds can be effective in connecting very different works together, although it must be noted that this can occur to the detriment of the album. Taking longer than necessary between tracks can also disconnect the flow of the album, so care should be taken to find a happy medium (and there will be one). Other reasons for doing this are to leave hidden tracks at the end of an album.	Sting's 'If I Ever Lose My Faith In You' from *Ten Summoner's Tales* has a very long fade out which is very gradual and adds a real essence to the flow of the album. Instead of a short gap after the long fade out, this gap is also long, some 8 seconds. This separates out the rock track into the second track of the album 'Love Is Stronger Than Justice' with its use of different time signatures and historical story. Another example of this is between tracks 1 and 2 on John Mayer's album *Born And Raised*. As with the Sting example above, track 1 ('Queen Of California') fades over a long time to silence and then has a long gap before track 2 ('The Age Of Worry') begins.

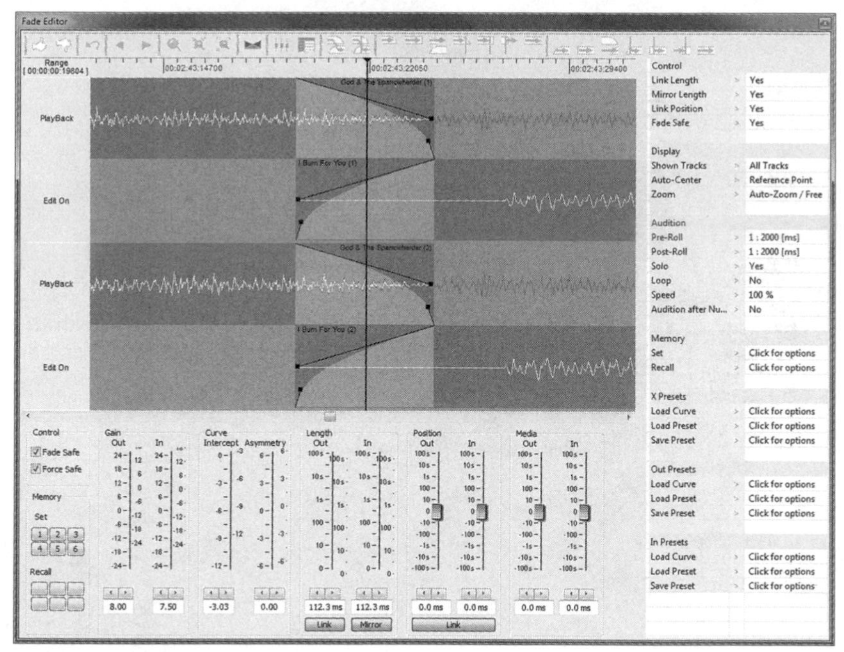

Crossfades can be highly tailored here in Pyramix.

What remains and, it could be argued, what is important, is the flow of an album. Again, this is subjective. How the album flows is possibly one of the most important things to keep in mind during all the micro aspects of the album through to this point. Going back and tweaking can be beneficial too until you arrive at a cohesive result.

8.5 Fades on tracks

One thing that used to set your average DAW apart from those adopted by the professional mastering world, such a Pyramix, Sadie, Sequoia and Sonic Solutions (or SoundBlade as its now called), was the way in which fades could be managed. Much domestic or semi-pro audio workstation software, usually intended for recording and mixing, tended to manage limited one-point fades. They'd simply permit you to choose the fade types, whether cosine, sine etc.

Historically, larger DAWs permitted you to have more control over your fades, thus allowing you to tailor aspects of the audio to the needs of both the music and the clients' wishes. Therefore, to suggest that fades can be prescribed in advance is somewhat ill-conceived. Fades should be considered with the appropriate amount of detail

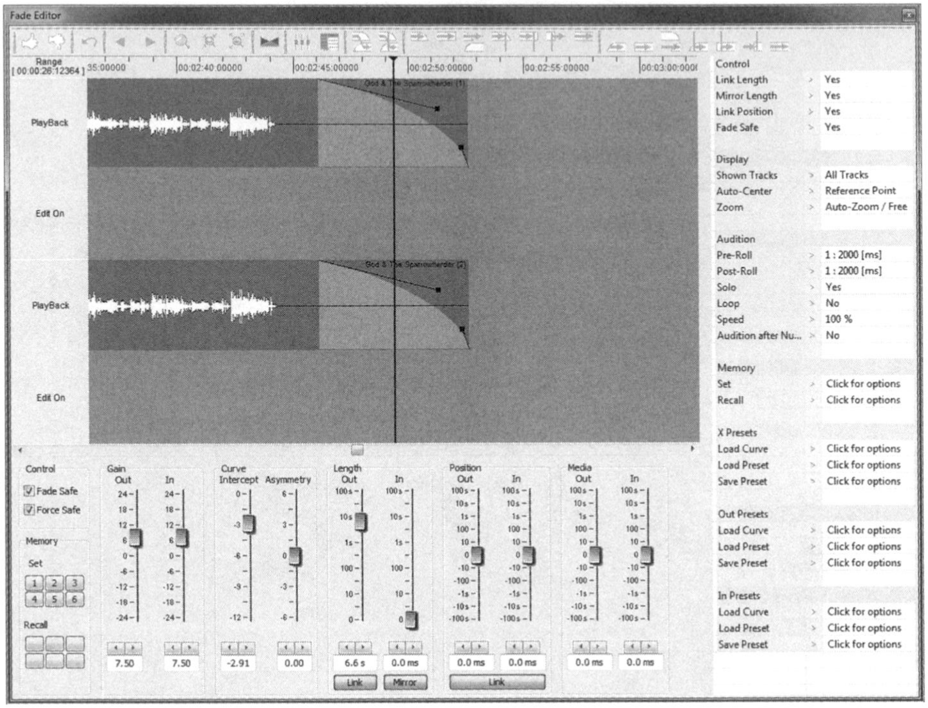

Fades can follow a variety of standard curves such as cosine, sine, or can be more precisely tailored.

and then rehearsed to see what works. Trial and error might be the name of the game.

Of course, some clients will be very happy with the fades and gaps you provide as this is a professional presentation of the songs they care about. However, some artists will have a strong view of how the whole album should feel, and segueing the tracks together will almost be as important to them as the individual songs themselves. Working intently with a client on this can produce some excellent results. Your advice might counter their expectations, but there should be no problem with trying out all the suggestions if time permits.

Fades are therefore not always as simple as a quick linear line from loud to quiet over a short period of time. Conversely, this could be just what the track needs. You will soon get a feel for how a fade should be handled. The considerations are similar to those we discussed regarding gaps in Section 8.3. To some degree the track will dictate the fade it requires, but so may the track that comes afterwards. Therefore the fade might need to be something that has a fast initial decline followed by a shallow fade to silence over many seconds. It is worth experimenting to see what provides the best 'feel' to the transition. Many DAW's, especially the professional ones, provide a range of preview controls that permit you to audition from the end of one track (and its fade) into the start of the next one.

The shape of your fade can never be prescribed, as it will be based on reactions to the instrumentation at the end of the piece and perhaps any level reduction the mix engineer may have instigated in the recording studio. You will be reacting to these factors and thus a fade may 'look' higgledy-piggledy to you on screen, but sound smooth and reactionary to the music.

8.6 Types of fades

Fades come in all shapes and sizes. In the Figure 8.4 there are four main types of fade. In mastering, you will probably find yourself editing these to be more precise, but starting with the linear fade it is possible to see what edits, or alternative selection, could be better employed.

Linear fades are best employed as a starting point. From here it is worth listening out for information in the quiet aspects that need increasing in level sooner. If this is the case, it is worth editing your fade to achieve a *logarithmic scale*, where the level increases quite rapidly. In

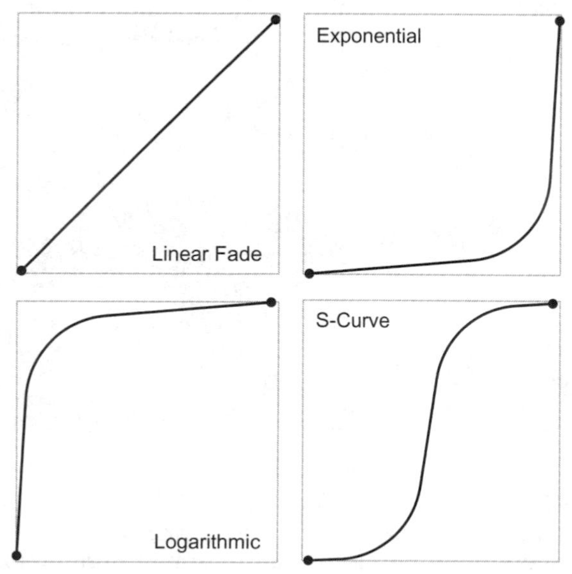

Fades can be tailored from the four main types shown here: Linear, Logarithmic, Exponential and the popular S-Curve.

the examples below this is a quite a vicious and steep increase. If your DAW permits it, tone this down to suit the audio accordingly.

S-Curves are quite popular, but again in our experience require editing in some way or another to enhance the impact of the fade. Sometimes it is necessary to increase the duration of the fade, and each system will have a different way of attacking this task – from a simple swipe with the mouse, to a fader that permits perfect honing as in the image from Pyramix below.

8.7 Segues

Segues describe the connection between two pieces of audio. It's a very common radio broadcasting term relating to the way in which two tracks might be mixed during a broadcast, with perhaps a broadcaster/presenter link in the middle. The same can happen in

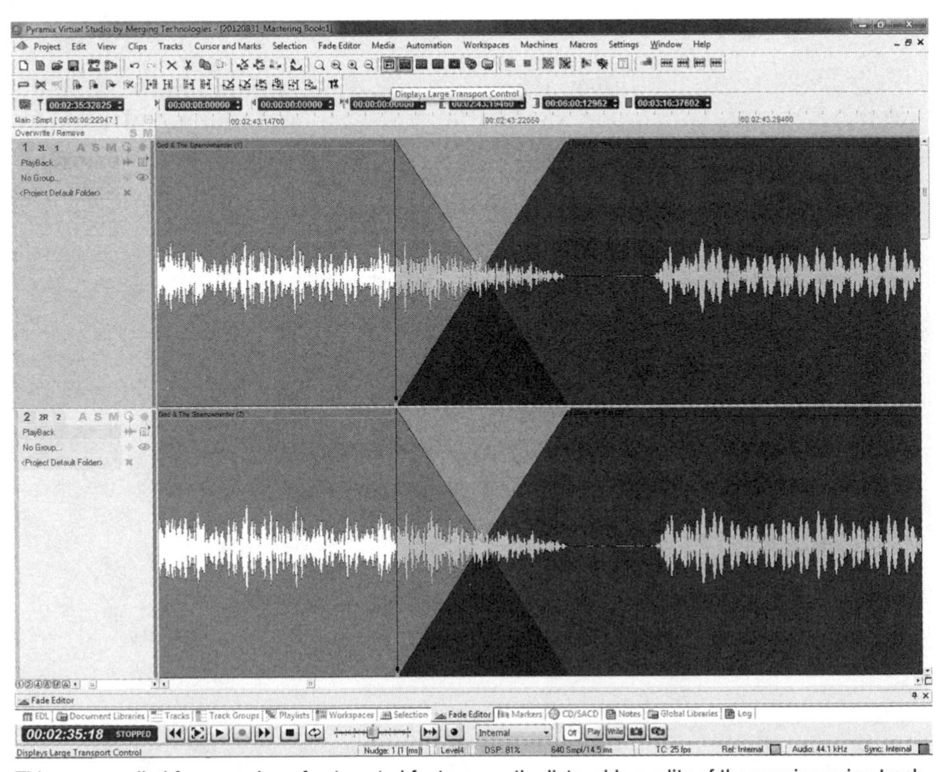

This segue called for a number of automated features as the listenable quality of the new incoming track was to appear 'older' than it actually was. The track begins with its bandwidth reduced (to meet the track previously) and then opens up in realtime to be a modern mix.

mastering (without the presenter link, but then again...). Some clients will want you to merge tracks together either using the two tracks as the segue, or by introducing a sound effect or other audio snippet to act as the glue between the two, especially if aiming for a more conceptual album. Many acts have done this pretty much throughout their careers, Pink Floyd being a key example.

Managing segues can be fun. There is often more than one dimension to consider, in as much as it is not just one track fading out as the next one fades in – although that's the simplest form. Other aspects, including more detailed level automation and the control of EQ's, Compressors etc., might be all called for to achieve the necessary segue.

8.8 Level automation

So far we've dealt with the fades of tracks at their start and end points, but on occasion it will be necessary to automate their level mid-way through. As we've already established, you're after what we've called flow across an album or EP, and in trying to achieve this it is important to maintain a level of 'quality' throughout. With this in mind, it may just be possible that a track waivers in level across its duration. You may therefore choose to reshape elements – between the verse and chorus, for example – before it reaches a compressor. In this instance, you may set up your processing to work perfectly with the verse, providing a really powerful and edgy entrance to the song, only for the chorus to then explode yet further in the loudness stakes causing pumping etc.

The Faderport is an inexpensive way to gain control over your audio levels with a real fader!

Knowledgebase

Red book preparations

As you will have gleaned from Chapter 3, there are several things that we should be aware of as we prepare our Red Book CD master. If you're mastering for a 'client', and especially if the artist cannot be with you for the session, you may wish to contact them and ask if they can complete what many studios call an 'engagement form'.

We use our engagement form (it could be called something slightly different by other studios) to ask for as much information from the client as they can you give us. This helps us to appreciate the order of the tracks, any codes that have been obtained, or need obtaining, such as ISRC and UPC/EAN codes (we'll cover these properly in the Chapter 9) and any personal information relating to the project that is needed to complete the mastering.

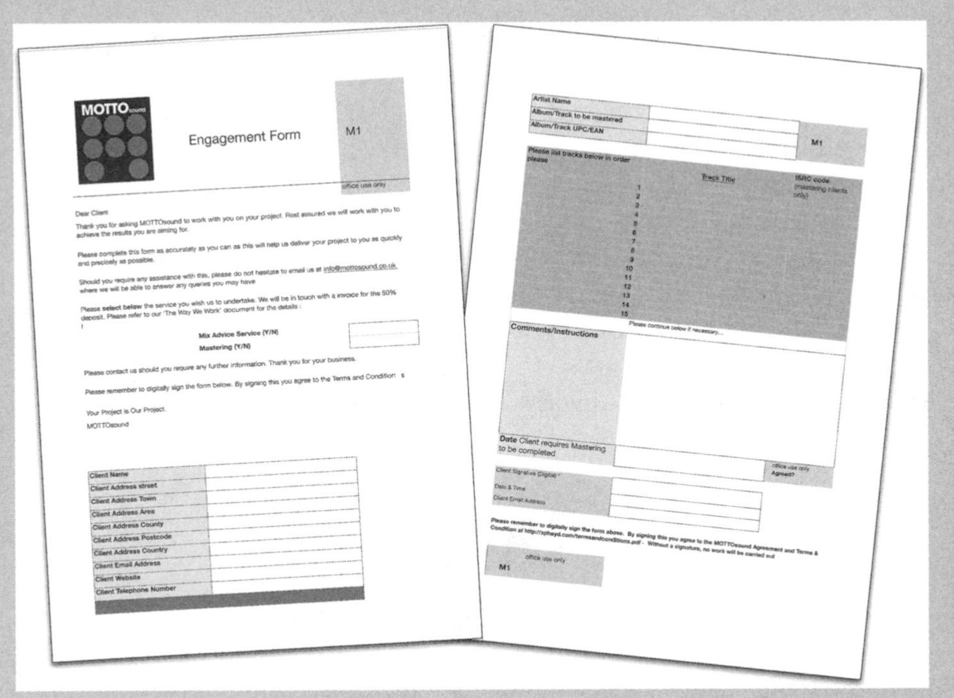

MOTTOsound

Engagement Form

M1

office use only

Dear Client

Thank you for asking MOTTOsound to work with you on your project. Rest assured we will work with you to achieve the results you are aiming for.

Please complete this form as accurately as you can as this will help us deliver your project to you as quickly and precisely as possible.

Should you require any assistance with this, please do not hesitate to email us at info@mottosound.co.uk where we will be able to answer any queries you may have

Please **select below** the service you wish us to undertake. We will be in touch with a invoice for the 50% deposit. Please refer to our 'The Way We Work' document for the details :

Mix Advice Service (Y/N)

Mastering (Y/N)

Please contact us should you require any further information. Thank you for your business.

Please remember to digitally sign the form below. By signing this you agree to the Terms and Condition s

Your Project is Our Project.

MOTTOsound

Client Name	
Client Address street	
Client Address Town	
Client Address Area	
Client Address County	
Client Address Postcode	
Client Address Country	
Client Email Address	
Client Website	
Client Telephone Number	

Artist Name	
Album/Track to be mastered	
Album/Track UPC/EAN	

M1

Please list tracks below in order please

	Track Title	ISRC code (mastering clients only)
1		
2		
3		
4		
5		
6		
7		
8		
9		
10		
11		
12		
13		
14		
15		

Please continue below if necessary...

Comments/Instructions

Date Client requires Mastering to be completed

office use only Agreed?

Client Signature (Digital)	
Date & Time	
Client Email Address	

Please remember to digitally sign the form above. By signing this you agree to the MOTTOsound Agreement and Terms & Condition at http://xphayd.com/termsandconditions.pdf - Without a signature, no work will be carried out

office use only M1

Perhaps a bit overkill if you're mastering at home, but if you're starting to master for others you might wish to obtain as much information from the artist/producers as possible well in advance of having to do the red book CD master.

One way you could deal with this is by using snapshot automation on a DAW, which we'll cover later (this can be emulated with analogue equipment, but not in real time), but the other is to automate the level of the audio file before it reaches your processors.

There are many ways to manage this, whether it is by employing your external controller surface, such as the FaderPort shown in Figure 8.6, and capture the automation live, or by drawing in the fades and points straight into the DAW using the mouse. This can be far more forensic when you need to edit at a detailed level. Most of our time is spent using this method, but might have been instigated by an initial fade on a controller surface.

Other level-automation might come in useful when evening out issues with the mix. A certain loud sound effect, or slightly quiet introduction, may need a shift in overall level. Depending on the DAW, this can either be drawn directly onto the waveform using its automation, or could be done manually using a fader with the 'write' automation button engaged.

8.9 Other automation and snapshots

Using a DAW allows you to do many things with automation that it was once not possible to achieve with the same accuracy. In the not too distant past, it was not possible to 'automate' your outboard processors to change a threshold setting on a compressor between the verse and chorus of a track. Instead, you'd master your track using each setting and perhaps splice up the master tape to switch between the two thresholds as you needed them.

The same technique is still used by many engineers using some of the highest-end analogue equipment. They will record the two versions into their DAW and edit them together in the same way. The alternative, if you're working completely in the digital domain, would be to use snapshots. Snapshots are 'images' of all your settings at any one time. With modern automation within the DAW it is possible to instantaneously recall these settings and switch back again as necessary. This can permit you to alter the EQ settings and perhaps the compressor settings immediately, as you pass from the quieter than necessary verse to the overly hyped chorus, evening the whole thing up.

Snapshotting is useful to assist in making quick changes between sections, or even tracks, as necessary, but it might be more prudent to make more graceful changes, such as a sweeping EQ that reflects some change within the music. DAWs again offer this functionality, just as in mix automation, and can be a saving grace. Naturally, it is quite possible to do some of these changes live if you're using analogue equipment, but most mastering engineers probably will work digitally for jobs such as this.

Automation does not stop just at EQs and snapshots. All DAWs now have the ability to dynamically alter pretty much any parameter, right down to the smallest control nestled in the compressor, in real time. It is therefore possible to sculpt complicated and detailed automation edits that fit with the music at hand, such as the example on segues in Section 8.6.

As with all things in mastering, it is often tempting to adopt these detailed techniques, as they provide power and control, but they should only be considered if the track calls for the change and you therefore feel it is necessary. It is quite possible that the track does not need such intervention.

8.10 Stem mastering

It is common to receive more than one mix of a track from the mixing session. This was borne out of the days before the total recall we enjoy with common pro DAWs such as Pro Tools, when setting up a studio and mixing console to recreate a mix might take two hours or more, and was therefore very expensive. To overcome any likely (and later expensive) issues in the mastering, it was common to produce more than one mix. The main mix would be the favourite everyone had chosen, but it would be typical for there to be a mix with the vocals up a little and another with the vocals down a little. Additionally, a mix with the vocals off would be provided if it was felt the artist would have to sing over the backing track, for example on the UK's Top Of The Pops BBC show.

This permitted the mastering engineer to select the most appropriate source track given the processing to be applied to it. For example, if multiband dynamics brought out the vocal too prominently it was possible to swap the files to the 'vocals down' mix and permit balance once again.

This has been expanded over recent years and appears to cloud the role of the mixer and the mastering engineer. Stems are essentially stereo-grouped files of the following components of a track: drums, basses, guitars, keys, vocals, backing vocals and percussion, plus any audio not listed here. These stereo files contain the unique balance created by the mix engineer, including all processing and effects, but just for the individual instruments. In the mastering studio, these stems are brought up in the DAW as though it was a mix session and lined up to the correct start point. This gives the mastering engineer a great deal of power more akin to that of a mix engineer.

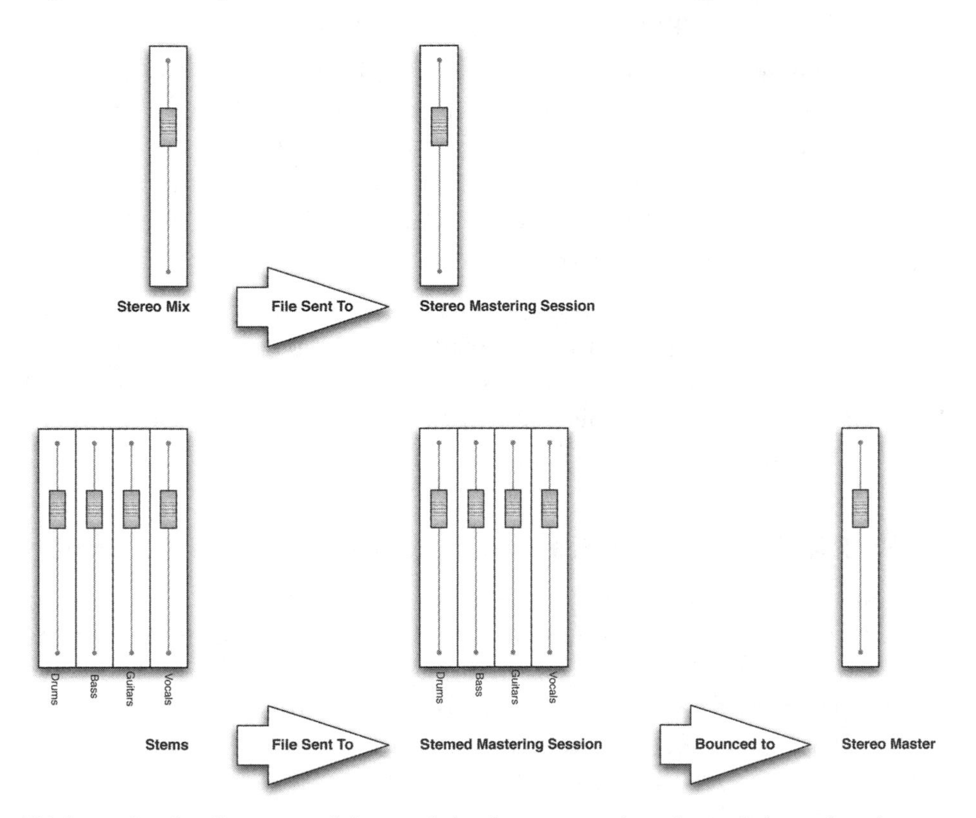

This image describes the process of stem mastering. As you can see here, the mastering engineer has many more decisions to make than in traditional stereo mastering. The role between mix and mastering engineer begin to get somewhat blurred.

Stem mastering can take a great deal longer to complete if you're being required to adjust many elements which you might have expected to have been repaired in the mix. The crux of the matter is that those decisions you'd expect to have been decided in the mix are deferred to a later point in the mastering room. Conversely, having this power at a later date in the mastering room can save time and hassle, should the main mix be perfect apart from one thing.

For example, say the chosen mix is very solid with all the instrumentation in the correct place. The only objection since mastering processing has been applied is that the vocals are too loud in the mix. In this instance, the mastering engineer has been provided with not just the solid stereo main mix, but stems of the track too. In this circumstance, the mastering engineer can line up the vocal stem alongside the main mix and invert the phase of the vocal stem. As this is inverted, the louder the fader of the vocal stem in the DAW, the less the vocals should appear in the output of the master.

As you can see from the above, stems can be very useful in addressing problems with individual instruments. It could simply be that the drums need to come up – if so, the drum stem can be added and turned up a little. So stems are not all bad, despite the negative connotations of leaving any difficult decision-making until the very last creative point in the production chain (i.e. mastering).

It is advisable for mix engineers to ensure that the main mix is as perfect as they can get it, and for security's sake to also bounce down the stems, as these can be extremely useful both in mastering, and from the musicians' perspective if they need to create backing tracks easily themselves at a later date.

8.11 Markers, track IDs and finishing the audio product

Up until this very point we have been dealing with the creative aspects of the mastering process – the more glamorous aspects of the game. However, there comes an important point when we believe we've achieved flow and the album is now feeling like one cohesive product. At this stage it's time we committed to the tracks becoming the album, and in doing so we need to prepare it for release.

In order to deliver the product we need to tell the DAW or other software which track is which. As we'll explore a little in the final chapter, different DAWs will have different ways to deal with this, but essentially you will need to tell your software where the CD track will begin and end and how to transmit this data to the final CD Master.

In Waveburner for example, this is a very graphical affair. You will place the audio tracks in the right place, and with a few natural edits here and there, the CD tracks will appear logically in place. On the whole this is fine, but you may wish to do some finer editing to achieve the exact results you want, with the markers that Waveburner provides. In the diagram below you can see the two tracks with a standard gap between them. Simply moving the tracks will provide immediate results regarding the gaps between tracks, but the pause sector can also be managed using the mauve flags.

Index points

As with all standards, there are simply excellent features that regrettably fall by the wayside. For some reason, we humans fail to see the good in some products or features (is VHS really better than Betamax – no it is not, but VHS won the war!). There are similarities with the *index point* in the CD format.

CDs are adorned with track IDs which we're familiar with given the discussion here, but there is also the option to utilize index points, although many CD players from the mid-1990s onwards do not include this option as they were never really employed properly.

Index points were intended for the listener to find specific parts within a CD track. For example, consider a classical compilation CD containing pieces from differing composers. Each composer's piece could be assigned a normal CD track ID. Meanwhile, each movement within each piece could be assigned an index ID.

This has been handled by either no longer providing a go-to marker (track ID or index point) for each movement, or for a long listing of track IDs to take the place of index points. As such, it is advisable for you to not use index points as they're largely unsupported, unless you wish to intentionally hide a track.

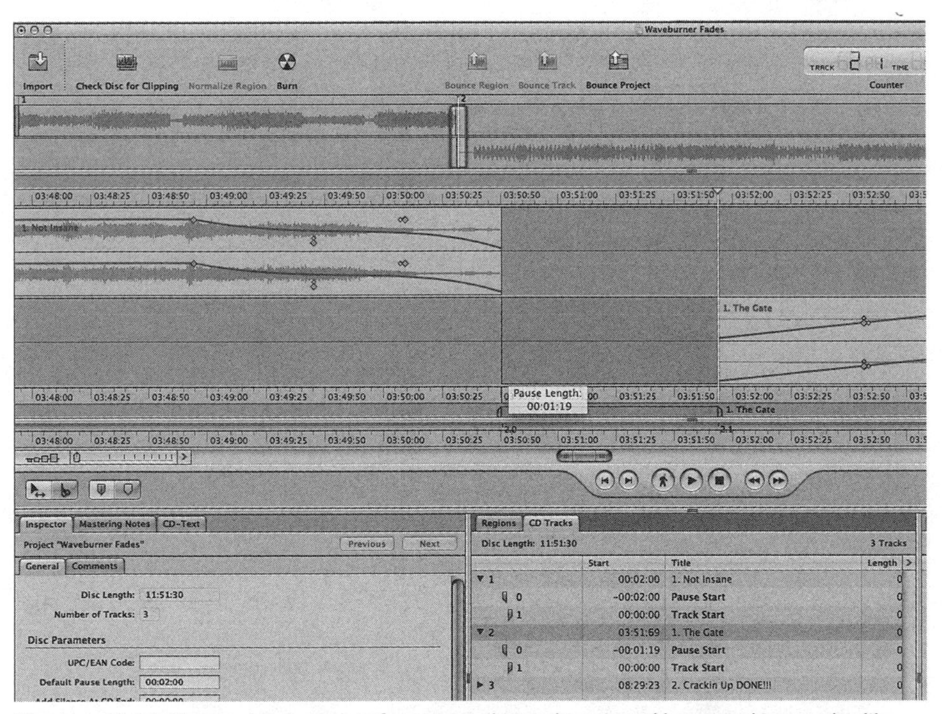

Waveburner's intuitive graphical user interface means that you're not mucking around too much with tables and markers like you are in Pyramix or other DAWs to manage your Red Book CD.

In other DAWs, in this case Pyramix, the CD is laid out using CD markers. This is more time-consuming to lay out to that of Waveburner, but provides complete flexibility to the mastering engineer when managing a project. In Pyramix, each start and end point is placed on the timeline as a marker. This, in turn, begins to list the tracks on the PQ pane where you will need to add the track titles and associated codes that we cover in the next chapter.

It is prudent to remember not to butt up the markers too close to the front of the track's audio, or in the case of Waveburner not to place the start of the audio too close to the start of the region. This is to permit older CD players and newer software players time to 'catch up', or 'rev up', especially if they have been skipped to that track! It has been known that fast transients can be cut slightly short when played through older players. Leave a very short gap before the audio track starts, just in case. This is advisable if your track has some silence, or air, before it kicks in.

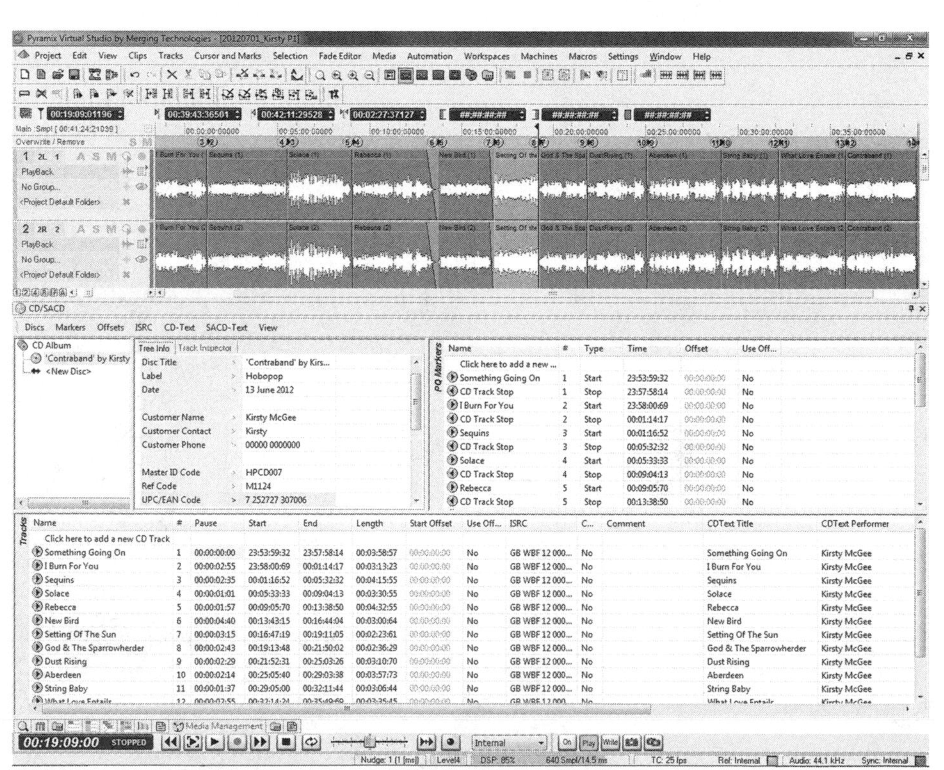

Pyramix uses dedicated CD Markers to lay out the track IDs for a red book master. Note the floating CD window with the track information in. This will need to be completed formally with the codes and details covered in the next chapter.

8.12 Hidden tracks

It has been trendy for some time to include hidden tracks on an album, although this seems to be fading now. Whether you can achieve this depends on the software you use and the type of hidden track you're trying to achieve.

Most hidden tracks are those at the very end of a CD, and are simply placed at the end of the last track with a large silent gap. So track 13 might actually be only a three-minute pop song, but will last seven minutes because the hidden track 14 has been added on the end. Other methods include placing the track in the sonic area before the start point of a track. The CD player, or computer, will seek the start point of the main song and ignore the gap before. In doing this, it is necessary for the listener to rewind on a CD player to locate the true hidden track. We assume hidden tracks are rarer these days (or perhaps they're not and we've not noticed because they're hidden!) as we've not been asked to hide tracks in the masters we've done lately.

With all the preparation outlined so far in the book, it is now time to explain how to deliver the product.

CHAPTER 9

Delivering a Product

In this chapter

9.1 Introduction 201
9.2 Preparation 202
9.3 International Standard Recording Code (ISRC) 202
Knowledgebase 204
9.4 The PQ sheet 205
9.5 Barcodes (UPC/EAN) 208
9.6 Catalog numbers 209
9.7 Other encoded information 209
9.8 CD-Text 211
9.9 Pre-emphasis 212
9.10 SCMS 212
9.11 Delivery file formats 214
9.12 DDPi and MD5 checksums 215
9.13 CDA as master? Duplication v. replication 216
9.14 Mastering for MP3 and AAC 218
9.15 Delivery 220
9.16 Final checks 223

9.1 Introduction

In Chapter 3 we introduced many of the concepts we'll talk about here in much more detail. This chapter is about the less glamorous stuff that goes with the job of being a mastering engineer. Every job has downsides – even the glamorous role of producer has boring and difficult parts to each session. We'd prefer the small inconvenience of delivering a product any day of the week – actually we quite enjoy it, as it's the culmination of a large body of work, from the artist's creation funnelled through the production chain to you! It's also a lot of responsibility to bear.

In the last chapter we brought everything we've covered in this book to a point where you can start to bring the mystery of mastering into focus. With the audio now in its right order, sounding like a flowing album, it's time to consider getting the product ready for market. Achieving this takes a little patience and preparation. Before you can begin to think about sending off a Red Book master, you'll need a series of codes, track names, catalog numbers and some general guidance from the artist, producer or label.

9.2 Preparation

Obtaining the aforementioned codes well in advance will assist the speed of the process. Unfortunately there a number of codes that you will probably want to apply for from various bodies and depending on which territory you're in, you'll may need to wait for a short period.

It is equally important that you get the relevant information from the artist, if that is not you, and the label, or whoever is 'releasing' the work. Items such as song titles (not working titles), album names, catalog numbers and CD-TEXT data (if different) all needs to be expressed to the mastering engineer in preparation for this stage.

If you're working with clients, it is sometimes convenient to consider presenting them with a form early on so that these questions are raised and are considered in plenty of time. Many an album has been delayed as certain codes or catalog numbers have yet to be issued.

9.3 International Standard Recording Code (ISRC)

The International Standard Recording Code (or ISRC for short) is the ISO3901:2001 standard, according to which all recorded pieces of music or sound should carry a unique identifier. This unique identifier is logged with the relevant rights society in each territory across the world.

The concept behind the ISRC is that if a track gains radio play at a station that is proactive in logging its playlist, the relevant fee will be paid to the writers based on play outs of your music. Unfortunately this is not always achieved, but many organizations take some pride in doing this, thus ensuring an income for the songwriters.

The ISRC system is managed by the International ISRC Agency at the International Federation of the Phonographic Industry

(IFPI) – www.ifpi.org/isrc – and there is an abundance of information about the code and various technical documents and bulletins that are worth reading if you're trying to understand more about this. However, the ISRC is fairly simple if you're being fed a code per track to assign to the music, as we'll explore below.

The UK's ISRCs are granted by PPL (repertoire@ppluk.com) and these codes can be sought for your recording. For the US it is RIAA (isrc@riaa.com). For all territories, please visit www.ifpi.org/content/section_resources/isrc_agencies.html.

As we are still considering the CD as the main vehicle for music delivery at the time of writing, the ISRC codes are encoded within the data on the disc. You might be interested to learn that, surprisingly perhaps, many MP3 files do not carry the ISRC codes, and this is often a consideration when working with digital downloads.

The ISRC, as per the ISO standard, is made up of a collection of digits containing unique identifier information that can be applied to every sound recording (see the ISRC breakdown box in the Knowledgebase). The code is unique to each recording, so if the same recording were to be placed on a compilation later on, it would keep the same code as the original release. However, if the track is a remix of the original, or a re-recorded release (not a re-release) then it would be given a new ISRC.

The ISRC begins with a country code, which is denoted by the IFPI and therefore managed by each territory that has been provided with a code. Next would usually be the registrant code, a three-character code made up of letters and/or numbers. This could be the producer, but could also be the songwriter themselves. Larger labels often manage these codes centrally and have a large range of them. In the example at the end of the chapter, given from IFPI, the label is Z03 – which is Mercury in France. This is then followed by the year (98 for 1998, 12 for 2012, and so on). Following this comes five digits that provide the unique identifier for the track itself. Each song, or recording, will be provided with a single code per track, and this will need to be logged with your ISRC holder for the territory you're in. In the UK's case, PPL.

The codes presume that one producer is not going to submit 99,999 tracks in any year, but were a label to do this one more producer code could be issued to accommodate it.

As Bob Katz is famous for saying, the mastering engineer's role is more about details than the snazzy aspects of making things sound great. We think he's right. There are many codes that need to be entered into the album (especially a CD), and logging these accurately has a direct impact on people's earnings. It is important that we take the handling of this data very seriously and ensure that things are checked and checked again so that everything is accurate.

Knowledgebase

ISRC breakdown

The ISRC consists of twelve characters representing country (2 characters), Registrant (3 characters), Year of Reference (2 digits) and Designation (5 digits). For visual presentation it is divided into four elements separated by hyphens and the letters ISRC should always precede an ISRC code. The hyphens are however not part of the ISRC.

(*The ISRC Handbook*, **www.ifpi.org/content/library/isrc_handbook.pdf**)

The ISRC Handbook also provides the following example of an ISRC code:

FR – Z03–98 – 00212
FR = France
Z03 = Registrant Code (3 characters and Z03 is Mercury France)
98 = Year of Reference (98 = 1998)
00212 = Designation Code (the individual code per sound recording)

Professional tools such as Pyramix have features to aid the handling of ISRCs. For example, Pyramix can incrementally insert codes from the last used, and if one label is mastering all its albums on one machine then these codes can be automatically generated per product and logged accordingly. So Client A could have GB ABC 12 00001 for the first track and then GB ABC 12 00012 for the last. Then Client B would start with the automatic code of GB ABC 12 00013. Similarly, if you are provided with sequential codes for the album you're mastering, then Pyramix can incrementally take your first ISRC and extend this for the remaining tracks. Each professional-level DAW will emulate or extend these sorts of features.

If you have to construct a compilation album, then you'll be placing unique ISRCs in for each track, and great care should be taken to ensure this is done correctly. It can get quite mind-boggling when you've got 50 tracks over a double-disc release!

9.4 The PQ sheet

The ISRC codes can and should be double-checked later on. It is imperative that your DAW system should be able to create something called a PQ sheet, which we'll come onto later, but what is PQ?

Without wishing to get too involved in the specification of the CD, this little bit is important to this chapter. Each stereo sample of data consists of a 16 bit word for the left-hand channel, a 16 bit word for the right-hand side. Alongside these lie some additional bits carrying information called the subcode. This carries vital information for the CD player about the music, and there is space for other subcodes as necessary,

The P part of the subcode is actually just one bit that changes state when the CD player needs to know there's a new track from a particular point onwards. The Q part denotes a subcode channel used to carry much of the information we submit into our PQ tables and what appears on the PQ sheet, hence the name. The Q subcode channel actually carries all the data detailed in the rest of this section. The R to W subcode channels store the later adoption of the CD-TEXT standard we'll discuss in more detail later.

The PQ sheet is a table of information that provides the pressing plant with an opportunity to triple check the audio presented to them against what the mastering engineer and the mastering DAW intended. Therefore all lengths of tracks, orders and gaps can be checked and any anomalies raised, before going ahead with the expensive pressings.

Most mastering software can produce PQ sheets, from the big professional systems all the way down to Logic's Waveburner and Wavelab. If you're likely to master using a DAW such as ProTools or Cubase, then other software will need to be employed to help manage this. One of the best is DDP Creator from Sonoris (www.sonorissoftware.com). As the name indicates, this creates DDPs from a number of formats, but most importantly will permit you to compile a Red Book master from a collection of audio tracks. DDP Creator is fully fledged and permits you to burn off a Red Book audio CD for auditioning or sending off for very small run duplications, but most importantly creates a DDP, which we explain later in more detail.

PQ Sheet

All times are in M:S:F format

Title	: Contraband	Engineer	: Russ Hepworth-Sawyer
Performer	: Kirsty McGee	Studio	: MOTTOsound
MCN	: 7252727307006	Phone	: 01138153001
Printed at	: Monday, July 30 2012, 08:45:15	Client	: Big Oak Songs
Created with	: Sonoris DDP Creator v3.0.1	Project	: Kirsty McGee - Contraband
Registered to	: Russ Hepworth-Sawyer - MOTTOsound	Source	: Digital

T	X	ISRC / TITLE	PERFORMER	START	LENGTH
01		GBWBF1200023			
	00	Pregap		00:00:00	00:02:00
	01	Something Going On	Kirsty McGee	00:02:00	03:58:57
				TOTAL	04:00:57
02		GBWBF1200024			
	00	Pregap		04:00:57	00:02:55
	01	I Burn For You	Kirsty McGee	04:03:37	03:13:23
				TOTAL	03:16:03
03		GBWBF1200025			
	00	Pregap		07:16:60	00:02:35
	01	Sequins	Kirsty McGee	07:19:20	04:15:55
				TOTAL	04:18:15
04		GBWBF1200026			
	00	Pregap		11:35:00	00:01:01
	01	Solace	Kirsty McGee	11:36:01	03:30:55
				TOTAL	03:31:56
05		GBWBF1200027			
	00	Pregap		15:06:56	00:01:57
	01	Rebecca	Kirsty McGee	15:08:38	04:32:55
				TOTAL	04:34:37
06		GBWBF1200028			
	00	Pregap		19:41:18	00:04:40
	01	New Bird	Kirsty McGee	19:45:58	03:00:64
				TOTAL	03:05:29
07		GBWBF1200029			
	00	Pregap		22:46:47	00:03:15
	01	Setting Of The Sun	Kirsty McGee	22:49:62	02:23:61
				TOTAL	02:27:01
08		GBWBF1200030			
	00	Pregap		25:13:48	00:02:43
	01	God & The Sparrowherder	Kirsty McGee	25:16:16	02:36:29
				TOTAL	02:38:72
09		GBWBF1200031			
	00	Pregap		27:52:45	00:01:64
	01	Dust Rising	Kirsty McGee	27:54:34	03:10:70
				TOTAL	03:12:59
10		GBWBF1200032			
	00	Pregap		31:05:29	00:02:14
	01	Aberdeen	Kirsty McGee	31:07:43	03:57:73
				TOTAL	04:00:12

Page 1

The PQ sheet is incredibly important in quality control. The mastering engineer must ensure the data is correct and get it widely checked by artist, producer and label equally long before the sheet gets to the manufacturing plant.

PQ Sheet

All times are in M:S:F format

Title	: Contraband	Engineer	: Russ Hepworth-Sawyer
Performer	: Kirsty McGee	Studio	: MOTTOsound
MCN	: 7252727307006	Phone	: 01138153001
Printed at	: Monday, July 30 2012, 08:45:15	Client	: Big Oak Songs
Created with	: Sonoris DDP Creator v3.0.1	Project	: Kirsty McGee - Contraband
Registered to	: Russ Hepworth-Sawyer - MOTTOsound	Source	: Digital

T	X	ISRC / TITLE	PERFORMER	START	LENGTH
11		GBWBF1200033			
	00	Pregap		35:05:41	00:01:37
	01	String Baby	Kirsty McGee	35:07:03	03:06:44
				TOTAL	03:08:06
12		GBWBF1200034			
	00	Pregap		38:13:47	00:02:55
	01	What Love Entails	Kirsty McGee	38:16:27	03:35:45
				TOTAL	03:38:25
13		GBWBF1200035			
	00	Pregap		41:51:72	00:01:61
	01	Contraband	Kirsty McGee	41:53:58	03:49:54
				TOTAL	03:51:40
14		GBWBF1200036			
	00	Pregap		45:43:37	00:02:28
	01	All Things Must Change	Kirsty McGee	45:45:65	02:27:63
				TOTAL	02:30:16
		LeadOut		48:13:53	

Page 2

Don't be too hasty with respect to codes, as it's an important aspect to get right. Ensure the client (whether that be the producer, other musician or indeed you) has sight of both the PQ sheet and the audio master to approve before you send it off. There have been many instances where fingers have been wagged inappropriately at mastering engineers for mistakes that have been introduced elsewhere. It is worth ensuring that the client has approved the master and accepts this formally so that the blame does not land on you.

9.5 Barcodes (UPC/EAN)

The other codes to embed within your CD are the UPC/EAN code – the Universal Product Code (also known as the Media Catalog Number – MCN) or the European Article Number. Essentially barcodes to you and I.

Each variety differs dependent on the territory you're in. Therefore the barcode should correspond on the graphics on the back of the

The barcode can be seen on the bottom right hand side of this compact disc back cover.

CD for the scanners in the store to gain the stock information and the price to charge you, but these codes can be entered into the data encoded on the disc too for identification purposes.

The concept behind the barcode is that every item has a unique identifier, which not only indicates which product is which, but which also performs some snazzy stock-taking and re-ordering functions for the retailer. Again, these are obtained by a licensee within the territory you're in. For the UK this is www.GS1UK.org. Within the US barcodes can be obtained from www.gs1us.org.

9.6 Catalog numbers

Other codes also exist within the CD project, but might not be considered as imperative as the barcode and the ISRCs. One such code is the catalog number provided by the label as a product code.

Labels will have a catalog number for each release and they're often a mix of letters and numbers and can be encoded within the CD. Labels often print this code on the spine of the CD case graphics.

The catalog number can also be encoded onto the CD in the same way as the ISRC and UPC/EAN codes. There are no hard and fast rules on this, but typically they can be anything from five digits to a more common eight-digit code such as LABEL123 and will be best encoded within the product and shown on the spine of the physical release. The code will also appear on the PQ sheet as another reference to double-triple check with cataloguing. Again, it is worth noting that the catalog number can actually be any string of numbers or letters and can be managed according to the rules of the label.

9.7 Other encoded information

Other information can be encoded on the disc such as the engineer, the producer and the studio. Currently (and regrettably) this information is not really being notated as it once was with 12-inch vinyl LPs. There are considerable moves afoot to encourage the encoding of much more information than is being currently captured. There are many campaigns brewing both here in the UK and in the US with respect to this.

Here in the UK, the Music Producers Guild has been running their Credit Where Credit Is Due campaign for some years now (http://www.creditisdue-mpg.co.uk). It has been highlighting the need to get album credits visible again. If you download an album from the Internet from a digital download store it is almost impossible to find out who played the second synth part in a particular song as the data is just not captured or stored. Similarly in the US, there are moves afoot with a campaign from NARAS (Grammy) called similarly 'Credit Is Due'. A Credit Is Due member called Count wrote a *TapeOP* (no. 89) article making one point very eloquently.

Quality control

Regrettably, in recent times there have been some instances whereby mastering engineers have got the blame for things going wrong in the production of physical releases such as CDs. This occurs because of poor quality control throughout the process, depending on the pressing plants and human error. No mastering engineer gets it right all the time, but it is hoped that the appropriate quality-control checks will be taken before the master leaves the studio.

It used to be common practice to receive test vinyl pressings from which the producer, engineer and mastering engineer might get the opportunity to either approve it or investigate further should something untoward present itself. This practice has been somewhat lost and too much faith is being placed on digital delivery.

As such, it is ever more important that as many precautions are taken as possible before you deliver your master to a pressing plant. We've spoken about getting your house in order by ensuring the audio details are correct on the PQ Sheet, as it is hoped that the manufacturer will do checks and balances to it should the audio information not line up with the PQ sheet you provided. It is actually quite refreshing when a plant seeks clarification, as you know they're doing their part in the quality control process.

However, we dare say this is not always the case and discs are pressed up in large quantities that do not have the same integrity as those that left the mastering room. So what can go wrong? We've already highlighted the problems in mastering from a Red Book CD and all the errors that a CD-R can incur in use and in transit. Data files are more robust, and employing the appropriate file formats such as DDPi with an MD5 checksum should ensure data integrity.

It is also worth noting that there are moves within the industry to seek out changes in quality control throughout the process. One champion of this is Ray Staff and the Music Producers Guild Mastering Group, which is working to set out guidelines for the industry to adopt. For more information see www.mpg.org.uk/news_stories/367.

Count likens the practice to that of the online movie providers and cinemas (theatres) chopping off the credits on all the films on offer. He notes there would be a significant set of legal moves to revert the practice. Is this not equal to that of the CD booklet and LP data that is all too often being lost, regarding the supporting music provided by professionals that helps to bring these to market?

What does this mean for the lowly mastering engineer? Well we may, in the future, have to start to look towards capturing and storing this information on some new portal for carrying metadata. We speak later on about Gracenote/CDDB, and this could be a place to expand to accept the additional data. It is presumed that many customers would be as interested in reading this information on their current players (iPhones, MP3 players etc.) as they once were with their 12-inch LPs. It is important to know who worked on an album, and to us it is most interesting to know who mastered the work. To have this information readily available would be highly desirable.

It is therefore in your interests as a mastering engineer to complete all the data you can, even though many labels and professionals are not that interested in ensuring that every field is populated.

9.8 CD-Text

Within a professional DAW you will have the ability to enter in CD-TEXT, which is an additional layer of information stored in the R/W subcodes on a CD. CD-TEXT can be displayed on the front panel of you CD player or car stereo as you listen to a CD.

CD-TEXT is not read by many of the aggregators, or even iTunes, when you rip a CD. Therefore, CD-TEXT is a bit of an optional extra and was not adopted with all players, which is unfortunate as this would go a small way to alleviate the Credit Where Credit Is Due campaigns.

CD-TEXT permits the mastering engineer to enter in information about the track title, artist, album name, composer, arranger, artist etc. This can enhance the product and was an extended part of the standard, as it were, long before the widespread adoption of the MP3 and our server-connected computer platforms. It almost beggars belief that this information has not been adopted and uploaded to the various databases (CDDB) which handle this kind of data.

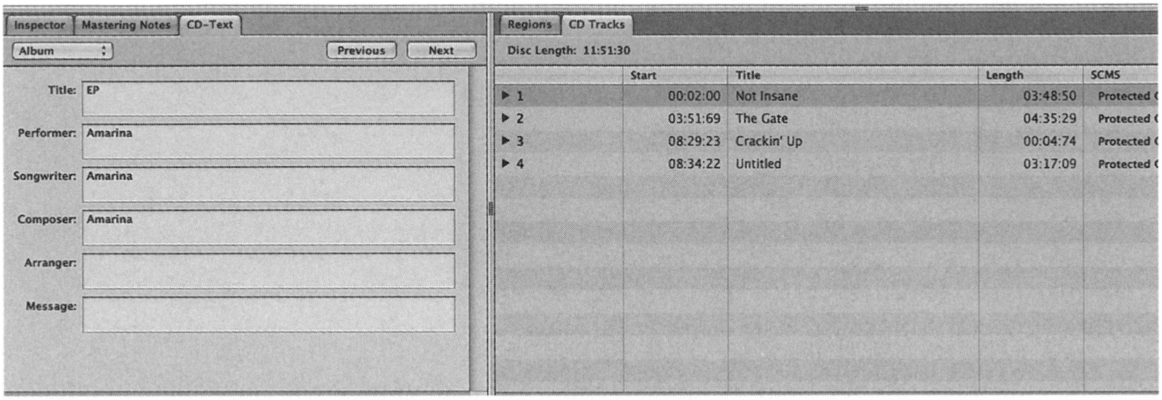

Inspector | Mastering Notes | CD-Text

Album | Previous | Next

Title: EP
Performer: Amarina
Songwriter: Amarina
Composer: Amarina
Arranger:
Message:

Regions | CD Tracks

Disc Length: 11:51:30

	Start	Title	Length	SCMS
▶ 1	00:02:00	Not Insane	03:48:50	Protected
▶ 2	03:51:69	The Gate	04:35:29	Protected
▶ 3	08:29:23	Crackin' Up	00:04:74	Protected
▶ 4	08:34:22	Untitled	03:17:09	Protected

Apple's Waveburner accepts data for CD-TEXT.

It is clear that an enhanced standard based around meta data will be needed for music as we go forward, as there are some definite moves to ensure this is catered for. CD-TEXT is placed in the subcode data of the compact disc and is entered into the workstation using tables usually incorporated into the main details pane. Below is an image from Waveburner with the CD-TEXT disc information to the left and the main CD Tracks pane to the right where the track title and other PQ information can be entered.

9.9 Pre-emphasis

In the early days of digital audio this facility gave a boost to the high frequencies (a little like Dolby B noise reduction in analogue). The lower-level high frequencies would be quantized accurately and therefore improved the quality of the system. The additional frequency boost would be attenuated accordingly upon decoding.

Naturally, this was conceived in the days prior to improved conversion techniques such as over-sampling and so on, but nevertheless the facility is still available and sometimes used for both audio broadcasts and in cutting records.

9.10 SCMS

The Serial Copy Management System (SCMS) was created as a response to the advent of Digital Audio Tape (DAT), a new format developed by a consortium of manufacturers intended to replace the compact cassette. The problem with DAT initially was that digital clones could be made of prerecorded, released compact discs, and objections from the record industry began to emerge. Not too long ago the 'Home Taping is Killing Music' campaign was prevalent throughout the industry. Little did they know that illegal downloading might bring the industry to its knees less than 20 years later.

During the 1980's there was a widespread campaign in the UK to discourage the copying of albums for further illegal distribution. How things have changed!

To meet with the complaints of the recording industry, SCMS was introduced which would prohibit the ability to digitally clone the material from one digital source to another if the flag was switched in the data stream.

Today it is perfectly possible to rip a CD using iTunes and more or less clone this using the .wav or .aiff file formats, and there's very little the recording industry can do to stop you. Efforts have been made to encourage Internet service providers to police this, but enforcing any such rule is proving to be very difficult. However, MP3 is currently the widely copied and shared format, and the record industry considers this as the lower-quality format that compact cassette once fulfilled. However, the piracy around this has far exceeded any early expectations.

The RIAA in America were keen to crack down on the sale of DAT machines, proposing a levy on the recordable media and hefty law suits for manufacturers releasing machines in the US. In doing so SCMS was permitted as a welcome response to the ability to clone CD recordings. However, the damage was done to the public image of DAT.

The whole prospect of DAT making it as a consumer machine failed, firstly due to the bad press that spread worldwide, but also because of the time it took to iron out the difficulties. As a result, the consumer, now hit by the inflated prices that RIAA and the record industry were requesting for the DAT tapes, failed to buy DAT in any meaningful numbers.

As a result there sat a high-quality digital audio recorder able to capture analogue audio at 44.1 kHz and at 16 bit – the same specification as the CD. In the early 1990s, digital mixing consoles were extremely rare and there were even fewer totally 'in-the-box' studios running on DAWs. As such, the DAT proved to be a relatively cheap and high-quality two-track mastering machine, replacing tired and expensive two-track 1/4-inch, and for those who could afford it, 1/2-inch open-reel machines that still needed digitizing to make it to the new distribution formal of CD.

DAT made inroads into the professional audio market very quickly and became a widespread standard for some 10–15 years until the proliferation of the CD-R and data transfer through the Internet. SCMS still remained, but many firms came up with devices that could be placed inline with the SPDIF data stream (we remember an HHB device) which could change the SCMS state in the data, permitting professionals to copy their own music (as the machines were no longer a threat to the cloning of CDs by this point).

Therefore, SCMS is something you still have control over when making your master CDs or DDPs, as can be seen in the table from Waveburner in Figure 9.4.

9.11 Delivery file formats

Mastering by today's standards is still, but only just, based around the compact disc, as this is a permanent and durable file format able to carry full-bandwidth PCM data. We have spoken previously in this book about mastering for vinyl, and there is some limited replication going on in this area.

We also recognize that many of you will be bypassing both of these physical mediums so will just be uploading to a distribution site. However, surprisingly some aggregators (the companies placing the files on iTunes and other download sites) will sometimes still request a CD. So preparing thoroughly for this medium is time well spent if working on an album or EP. Additionally, most mastering software is based around delivering to a Red book compliant format such as DDP.

Back in the early days of CD mastering, delivery was often difficult as there needed to be a portable recordable format to deliver the digital audio to the pressing plant. These were the days before recordable CDs, or even hard drives as we know them today. As such, the U-MATIC video machine could be used to encode digital audio with the appropriate convertor. The Sony PCM1610, and later 1630, was used widely to encode the audio onto a portable format which could then be transported to the pressing plant. Here's an example of when the PQ sheet would be imperative in ensuring the accuracy of the end product. The pressing plants would have spent more time ensuring the data lined up than perhaps is thought necessary today with current delivery formats.

9.12 DDPi and MD5 checksums

As time moved on, Doug Carson & Associates produced something called the Disc Description Protocol, or DDP for short (the image file is often referred to as the DDPi), which permitted engineers to take an image of the CD as one long audio stream. Within the data folder was a file explaining the PQ information, plus a 'checksum' to match the data integrity should any errors be introduced. If the checksum was incorrect, the data was rejected.

In the early days, Exabyte tape was used to send DDP information to the pressing plant. Originally Exabyte was employed by the IT industry to back-up servers and large quantities of data, and was therefore the perfect medium.

In the early days of CDs, we must remember that 650 MB was an astonishingly large amount of data to store. Hard drives with less capacity were incredibly expensive in the mid-1980s, costing as much as a family home here in the UK. These days 650 MB is considered, tiny and even our mobile phones carry much more than this.

Later, DVD-Rs and CD-Rs could be employed for the purpose of storing the DDP as a data file. This is a fairly robust method, but the Internet has provided the most cost-effective and efficient way of getting mastered data to the appropriate destination.

Now the DDP is often accompanied by an MD5 checksum, which ensures data integrity for the whole DDP folder, not just the DDP checksum for the audio image. The MD5 is a more robust checksum and it is good practice to ensure that additional data is not lost in transfer over the Internet. Many applications include this as standard, but many freeware applications exist to calculate your MD5 checksum before sending a DDP folder off for replication.

9.13 CDA as master? Duplication v. replication

On occasion we have been asked to provide a CDA master, or a Red Book CD master, from which a pressing plant will make a glass master. To put this into context we must first understand some of the economies of scale for the pressing plant.

To create a glass master, which will press the information into a regular CD you buy in the shops, costs some considerable money. There's a rule of thumb within the industry which suggests that if you're making 500 discs or more then it is cost-effective to prepare a glass master. If you're doing less, especially runs as low as 50 or 100 discs, then the costs don't quite stack up. Certainly, the lower you go below 500, the more advisable it is to duplicate CD-R copies from the original. In the latter case, it is quite conceivable that a verified Red Book master might be the only option on the table, as these machines take one disc in the 'top' machine in the rack of players and may clone five or more discs in real time.

However, when a glass master is going to be produced it is not advisable to take this information from a Red Book CD, despite the many requests from replicators that we've seen! A DDP should be insisted upon, as even verified Red Book CDs can introduce errors through dust, scratches, fingerprints and so on.

Verified CD masters are those that have been checked for errors and are within tolerances to the original copy. For many years we would

CD Duplicators are employed for smaller runs where it is not cost effective to prepare a glass master.

make masters using our Plextor drives and employ the accompanying PlexTools software to run an error check on the CD we'd just burnt. C1 and C2 errors are the ones we watch out for, and it's important that these are within the appropriate margins of the factory if they are to be used as the masters. For those replicators insisting upon this method, we'd hope that their quality control mechanism checks for data integrity.

DDP offers a low cost, high success rate, to ensure that data reaches the recipient in an accurate state. Verified Red Book masters are less advisable. Other options include sending your .aiff and .wav files and, upon your instruction, permitting the pressing plant to make up the discs for you (use a checksum such as the MD5 too). However, this may circumnavigate your ability to control the gaps and the levels you would wish to achieve as the mastering engineer.

Other formats exist such as PMCD (Pre Master CD) on the Sonic Solutions (now SoundBlade) system, which was intended to replace the Sony 1630 system but is now being replaced by the more widely adopted DDP.

9.14 Mastering for MP3 and AAC

For some mastering engineers it has long been considered best practice to master an album for the traditional medium of CD, and master the album slightly differently with the MP3, the MP4 and digital distribution in mind. At the time of writing, there are indications that the CD might be soon be a thing of the past, as labels consider finishing its stronghold in light of ever-rising digital downloads.

Unfortunately, for far too many years now some aggregators (the companies who place the audio information on the digital download sites) have taken a CD and ripped it using similar software to that of iTunes. For the same reasons as not wishing to duplicate from a Red Book master, issues could occur in a high-speed (64x speed) rip of the album.

Therefore, many mastering engineers have spent a great deal of time remastering the material they have prepared for CD to provide the best possible image of their client's work for the data-compressed digital download market. The mastering engineer spends the time necessary to ensure that the data reduction codecs do not spoil

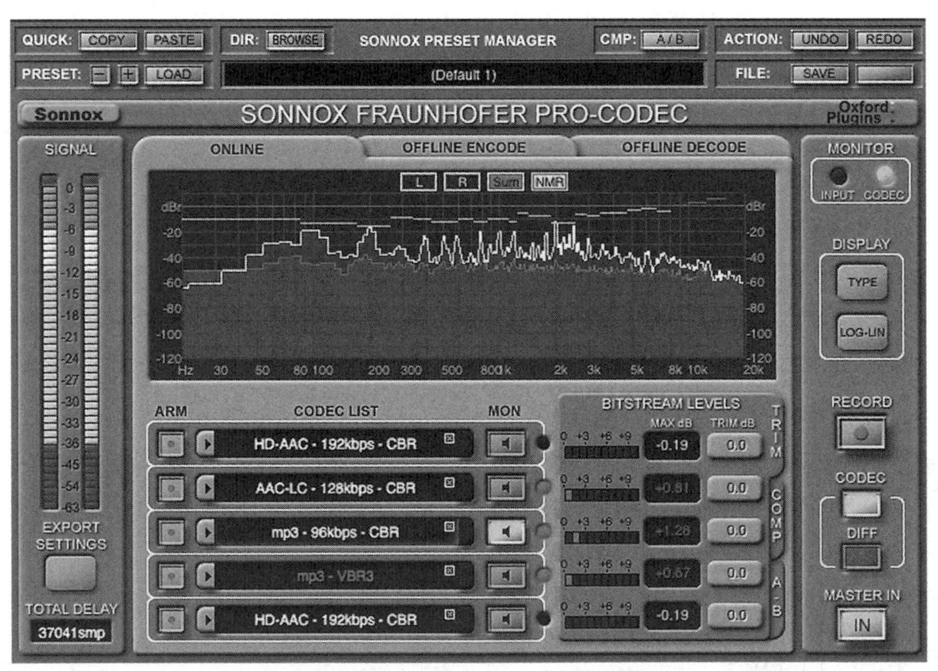

Sonnox's Fraunhofer Pro-Codec is a modern treat saving time for all mastering engineers who wish to create specific data compressed masters for their clients.

The future of data compressed audio?

Whilst data compressed formats were not necessarily developed out of a need to compress audio for the music market, they were successfully developed as part of the MPEG format to compress audio for video. The MPEG standard has become the most popular of all data compression formats for audio, and is the industry standard.

The MP3 and MP4 (AAC) standards are well established for the digital delivery of music over the Internet, and despite the relatively good quality for the much smaller file size, there are still many reservations about these codecs. These debates will rage on, no doubt, but we ought to ask how long we'll need to listen to data compressed formats?

Indeed, data compression codecs are widely accepted and entrenched in all our devices. Presuming that CD-standard audio quality (44.1 kHz and 16 bit, referred here to as PCM) remains a current uncompressed standard, noting that DVD-A and SACD have not made the impact we'd hoped, could our storage expand to allow us to carry PCM quality audio around with us instead? Or will we be encouraged to carry larger libraries of music as AACs?

It is heartening to consider that many consumers are now investing in expensive and sonically improved large-form headphones as they now listen to music predominately on the move. Perhaps this could lead to a renaissance with respect to high fidelity audio? In doing so, it will not be long before storage will meet with the needs of a keeping PCM, or lossless data, on our portable devices. This could also present a glimmer of hope that a discussion about the loudness wars could take place as the consumer reaches again for high fidelity audio.

Additionally, it is hoped that the consumers will join the producers in placing pressure on the music industry so that higher-quality audio is consistently made available to them via digital download sites. One benefit would be a reduced need for two versions of the master, one for the data compressed format and the other for the CD.

This is already being exercised in the longer-term concept behind the Mastered For iTunes campaign which, despite the current conversion to lower file sizes with AAC, will at least allow you to submit a higher-resolution file without lowering the standard to CD at 44.1 kHz and 16 bit.

the delicate balance of the audio material. These masters could be improved upon as the mastering engineer understands the concept of auditory masking, enhancing or lowering certain areas to ensure the best sonic representation for that format.

Some aggregators have resisted this in the past, and have not widely accepted the direct data delivery of compressed audio files for upload. However, this is starting to change in light of the potential, and quickly anticipated, demise of the CD.

To many this may seem an unnecessary burden to the production process, but it is no different to an alternative master being produced for the vinyl copy. This is just another specialized master engaged for the format it will end up on (perhaps an MP4 on portable headphones).

A heaven-sent plug-in has emerged from some Music Producers Guild members in conversation with Fraunhofer and Sonnox. The result was Sonnox's Fraunhofer Pro-Codec plug-in that permits real-time auditioning of your full-quality material in varying popular formats such as MP3, MP4, etc. This has enabled mastering engineers not only to tailor specialist MP3 masters of their work in real time, but also given them the ability to batch process into various data-compressed formats.

To some, there is a light at the end of the tunnel in the form of Apple's Mastered for iTunes. More details on this can be seen in the Knowledgebase, but there is some future proofing available going forward as Apple encourage submission of music as 96 kHz and 24 bit for future encoding or distribution.

9.15 Delivery

Getting your finished master to the pressing plant, or to market, can now take various forms as we've already explored within this chapter. Only a handful of years ago it would have been good old snail-mail – posting a verified Red Book CD master to someone for duplication, or a DVD-R of a DDP to a pressing plant. How quickly things have changed – we now post almost no CD-Rs at all.

Predominately, the delivery mechanism you use will depend on your client or output destination, but it is almost certainly going to involve sending something over the Internet. Preparing for this is not as simple as you might think if you've never done it before.

Despite the large download speeds we're all quoted to have in our Internet connections at home or work, the upload speeds are something entirely different. 650 MB might only take a matter of minutes to download, whereas the same could take anything in excess of one hour to upload. Preparing the necessary time for this is important if sending a larger album.

FTP clients such as Fetch can make the experience of FTP very easy - just like working in the operating system itself www.fetchsoftworks.com/fetch.

How to send this album is also a consideration. Email is obviously out as it hardly ever supports large audio files such as these. For the initiated, something called File Transfer Protocol (FTP) has been employed for many years, whereby the sender can use an FTP Client (a piece of software on your computer showing and allowing you to interact with a server's file structure) to upload large amounts of data. This is often a very good way of sharing large files and many pressing plants still accept DDPs in this way.

Labels and others have used the DigiDelivery format, which was sadly discontinued late in 2011. In our experience labels have been using some of the newer large-file transfer services that have come to the fore. Systems such as WeTransfer.com and YouSendIt.com come with some interesting client programs that allow you to send the appropriate file with receipts and additional passwords.

These, we have found, can be very useful and cost effective. An FTP server is usually sold as part of a website package (and URL) and can be more expensive.

Mastered for iTunes

At the time of writing, 'Mastered for iTunes' has recently been announced to the mastering world by Apple. Apple describes digital downloads as the 'dominant medium for consuming music' over the CD and has introduced this new method to encourage masters to be uniquely created for the iTunes platform, brandishing 'Mastered for iTunes' within its store.

On the face of it, Mastered for iTunes promises a way in which the mastering engineer can locally preview their high-resolution masters (usually above 16 bit and 44.1 kHz) in iTunes, plus AAC files (256 k bits-per-second variable-bit-rate AAC files), for checking the outcome once their files get on the store for sale. Mastered for iTunes is a scheme whereby the mastering house will be encouraged to submit to Apple a 96 k and 24 bit file of the music they're working on for two reasons.

Firstly this proposal overcomes the multiple data reduction processes that might occur. Starting with your 96 k, 24 bit master, you will dither this down to 16 and concurrently downsample this to 44.1 kHz. Usually this will then be transferred to CD and distributed to the customer. Very few listeners carry CD players around with them any more and are therefore likely to encode their discs within iTunes. They may have selected something like Apple's Lossless codec to achieve this, but more likely than not this will be set to encode the audio as AAC at perhaps 192 k. Therefore the audio is subjected to another process before it is listened to on the portable equipment. However, it is more likely that iTunes has been loaded with audio ripped from that CD in the first place which is more widely listened to. In theory, Mastered for iTunes attempts to overcome this additional process permitting you, within your mastering room, to audition how it will sound using the iTunes Plus 'Master for iTunes Droplet' application provided as part of the Mastered for iTunes initiative.

Secondly, there is a good deal of potential future-proofing here. Mastered for iTunes as a scheme recognizes that the future of audio will, and should, exceed 16 bit and 44.1 kHz, and as such Apple would like to posses these files locally so that they can instantly be re-processed in line with future technological and social developments in music consumption. This can only be a good thing given the longer-term aims, and the comprehensive repository that iTunes could then become.

So what is Mastered for iTunes? To the casual user it's just a download with a few very useful 'droplets' in it. However, unpacking this, these droplets are just useful little applications. The Mastered for iTunes droplet permits you to place a high-resolution audio file onto it and the droplet will transform your file into an AAC iTunes Plus file, which is exactly what will currently happen if you send iTunes a high-resolution file now – it will be transformed into this. The other droplet is also quite interesting as it converts your audio files in to .wav format. So any data in any format, compressed or otherwise, can be converted into the instantly transmittable .wav format. These are very useful tools regardless of the Mastering for iTunes scheme.

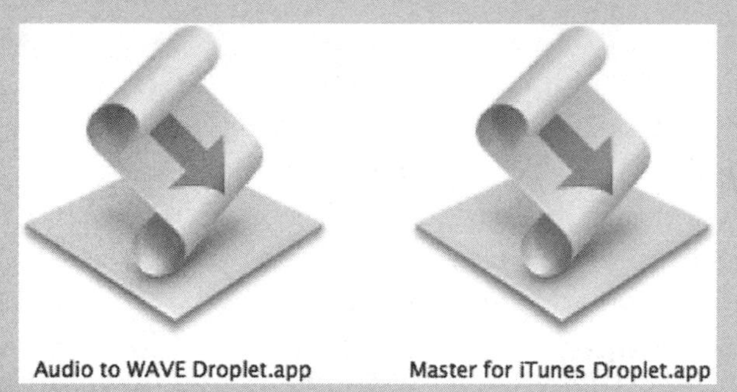

Apple's new Mastered for iTunes provides some interesting little apps called droplets permitting a quick format change of an audio file.

Perhaps the most important addition is the AURoundTripAAC Audio Unit, which gives you the ability to hear your full, uncompressed audio, and with a flick of the switch hear the Mastered For iTunes format sound. This will, just like the Sonnox Fraunhofer Pro-Codec plug-in, permit you to make necessary adjustments with your masters to create the best possible outcome. The regrettable thing is that these are currently only for the Apple Mac platform. Most professional mastering platforms are windows based.

There are numerous thoughts and comments across the Internet in relation to Mastered for iTunes, much of it not as positive as it could perhaps be. Reading between the lines, and given some of the knowledgeable discussion about this. It is perfectly easy to see that Apple appear to be future-proofing in a reasonably intelligent way. There are however just concerns that the Mastered For iTunes codecs don't do as good a job making lower quality masters as a mastering engineer might. It is a shame that metadata was not put on the agenda with Mastered for iTunes too!

Mastered for iTunes: www.apple.com/itunes/mastered-for-itunes/

Aspera, DigiDelivery's owner, offers new products for assistance with managed data transfer, but at the time of writing we're unsure as to how many DigiDelivery users have reinvested in such a replacement.

9.16 Final checks

With your CD now on its way to manufacture and soon to be pressed, it is worth turning your attention to wrapping up the project. It is always worth making some notes on the work you've completed, whether that's something unique in processing you had to try for the first time, or some technical gremlins you had to wrestle with. Understanding what you have done is extremely important as the memory can only hold so much.

How you manage this information is up to you, but it is advised that you manage this somewhere where it can be searched and accessed quickly. Many studios use databases built for the purpose, whether that be a programme such as StudioSuite by AlterMedia (www.studiosuite.com) or a more generic client record management system such as Daylite (www.marketcircle.com). Many studios we know have adapted the Filemaker database solution to permit them to keep up-to-date records on their clients and their project notes. Even a folder with some scribbled notes would do for the less technical (or Bento, which is proving popular on engineers' iPads!).

Presuming you've not been graced with a test pressing, and the CD has been pressed (or online variant released), it is worth obtaining a release copy and listening to check the audio is as good as when it left you. One simple way to check this is to pull up the DAW session file of the master and import the CD audio into it on a new track (this will not work properly for anything data-compressed). A surefire way to check would be to turn this CD audio out of phase and play it back. If you hear anything at all, you'll know there are differences and might choose to look further as to the reasons why.

Obviously, with downloaded versions this technique is less of a clone, but more of a processed copy, so listening will be the best test of quality control. Nevertheless, you should listen to the replicated/distributed versions to check for issues that might help you when visiting a similar project in the future.

By now you're equipped with the core knowledge necessary for pulling a professional release together. For more information about DAW-related differences, please turn to Chapter 10 for assistance.

Gracenote and CDDB

Gracenote, originally called CDDB (CD Database), is a database that holds all the song information usually gathered for recordings on iTunes and other players. As you pop a CD into the machine the software will use a connection on the Internet to access Gracenote and download the data relating to that song or album.

CHAPTER 10

DAW Workflow

In this chapter

10.1 Introduction 225
10.2 Method 1: Track-based mastering in a conventional DAW 226
Importing audio 226
Songs to tracks 226
Instantiating mastering plug-ins 229
Rendering pre-master files and applying fades 230
Final delivery 231
10.3 Method 2: Stereo buss mastering in an unconventional DAW.......... 232
MClass Equalizer 233
MClass Compressor 233
MClass Stereo Imager... 234
MClass Maximizer.. 235
A Word of Caution 235

10.1 Introduction

Having explored a variety of mastering techniques, now comes the inevitable task of trying to apply these processes in your DAW. As the last couple of chapters concentrated on the process of assembling a master in a professional Red Book-compliant application, we thought it was important to illustrate some of these principle techniques in a conventional music-based DAW. Of course, you can't necessarily expect to produce final DDP master file, but for a large part of the mastering process an off-the-shelf DAW should be reasonably sufficient for what you need to achieve.

What we're going to explore here are both the methods and practices behind a range of conventional music-based DAWs – applications such as Cubase or Pro Tools, for example–as well as some of the alternative strategies that need to be adopted with other music applications, such as Propellerhead's Reason. Although each situation is different, the broad principles are the same across a large variety of music applications and, most importantly, the overarching principles and practices of mastering remain the same whatever application you choose to work with.

10.2 Method 1: Track-based mastering in a conventional DAW

Most conventional DAWs follow the same working methodology, largely based on a traditional mixer interface in one window combined with a horizontal series of track lanes in another. Principally speaking, this approach is designed as a means of recording and mixing multi-track compositions, rather than handling a series of two-track master files. However, with a little refinement a DAW can be repurposed for mastering activities.

What is important to note with the principal DAW-based mastering workflow is that it's reasonably generic across a range of applications – whether it's Pro Tools, Cubase, Logic Pro or Sonar (to name but a few). In short, what we describe in the next few sections can largely be transferred across DAWs, and indeed, we've chosen to illustrate each step using a different DAW rather than tie it into a specific application.

Importing audio

Start by importing all the appropriate final mix files into your DAW, ideally without any form of compression, limiting or EQ having been applied when the files were initially created. You'll also need to establish an appropriate bit depth and resolution for your session, as some of the songs might need to be converted in order to import them into your session.

Songs to tracks

Deciding on how you're going to deal with a number of songs forms one of the key issues of a DAW-based mastering workflow, and there are a number of different solutions. Professional mastering DAW users are likely to place every song onto the same stereo track and

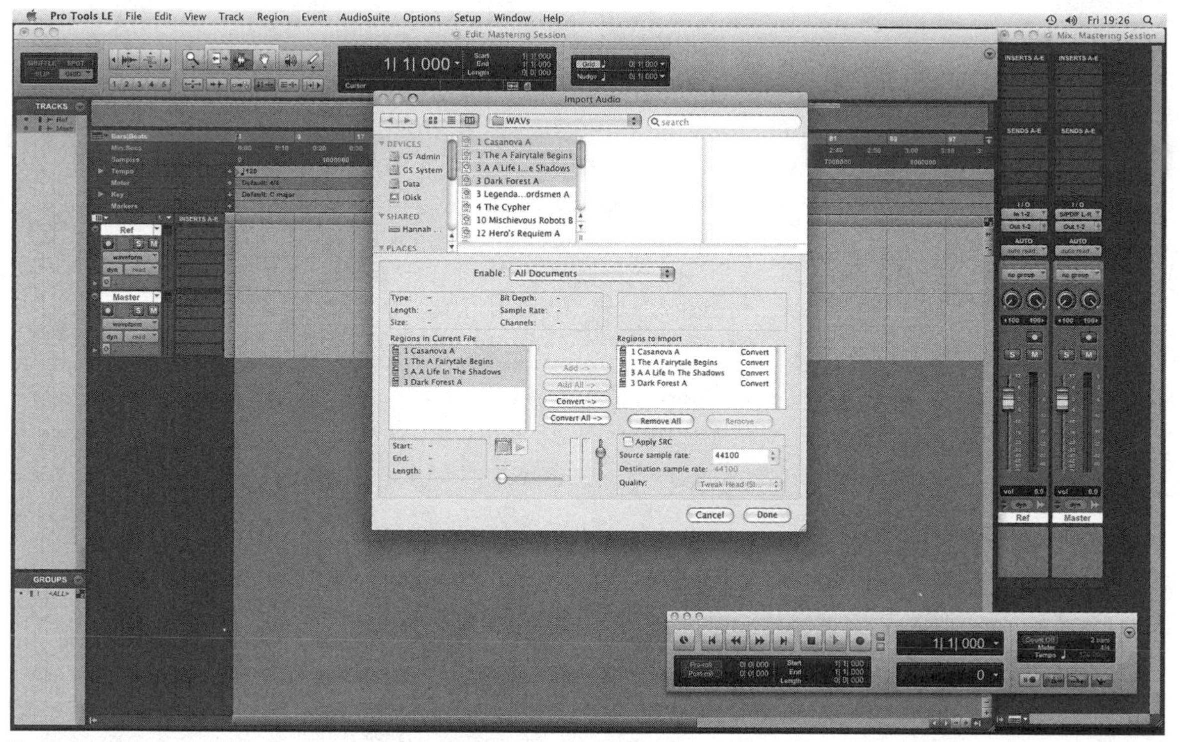

The final mixes all need to be imported into your final session, arguably applying any appropriate bit-rate or frequency conversion as part of this process.

employ snapshot automation (if necessary) to alter the processing. In the more popular DAWs we find the best approach is to use a separate track for each discrete song as part of your project. The principle advantage of this approach is that you can audition settings between songs quickly and easily and adjust them on a track-by-track basis. Ideally, of course, you're trying to create some degree of homogeneity between the different songs, so the ability to tweak any song at any point of the mastering process is a distinct advantage.

As well as placing your own songs on discrete tracks, you can also use one or more track lane for references masters, potentially sourced from a commercial CD. The significant benefit here is that you can apply some degree of loudness correction, arguably applying a small amount of level attenuation to account for the extra loudness the commercial CD will inherently have. Of course, as you add limiting to your master you'll might need to rebalance this correction so that you retain the uniform reference point between tracks.

Possibly the only principle disadvantage of a track-per-song system is the amount of processor power that you have at your disposal. If there are a lot of songs on the CD, for example, and lots of DSP-hungry up-sampling plug-ins used on each track, then you might find that you soon eat up your computer's resources without being able to process all the songs. In this case you always have the option of bypassing plug-ins temporarily and maybe working on half a CD at a time. Likewise, you could always choose the render each track, or bunches of tracks, as you work through the project.

As you assemble songs onto their relevant track lanes, you might want to consider applying some basic housekeeping with respect to edits at the start and ends of the songs. However, you need to steer clear of applying any lengthy fade-ins or fade-outs at this point, as these should only be applied after your mastering processing (including compression, EQ and limiting).

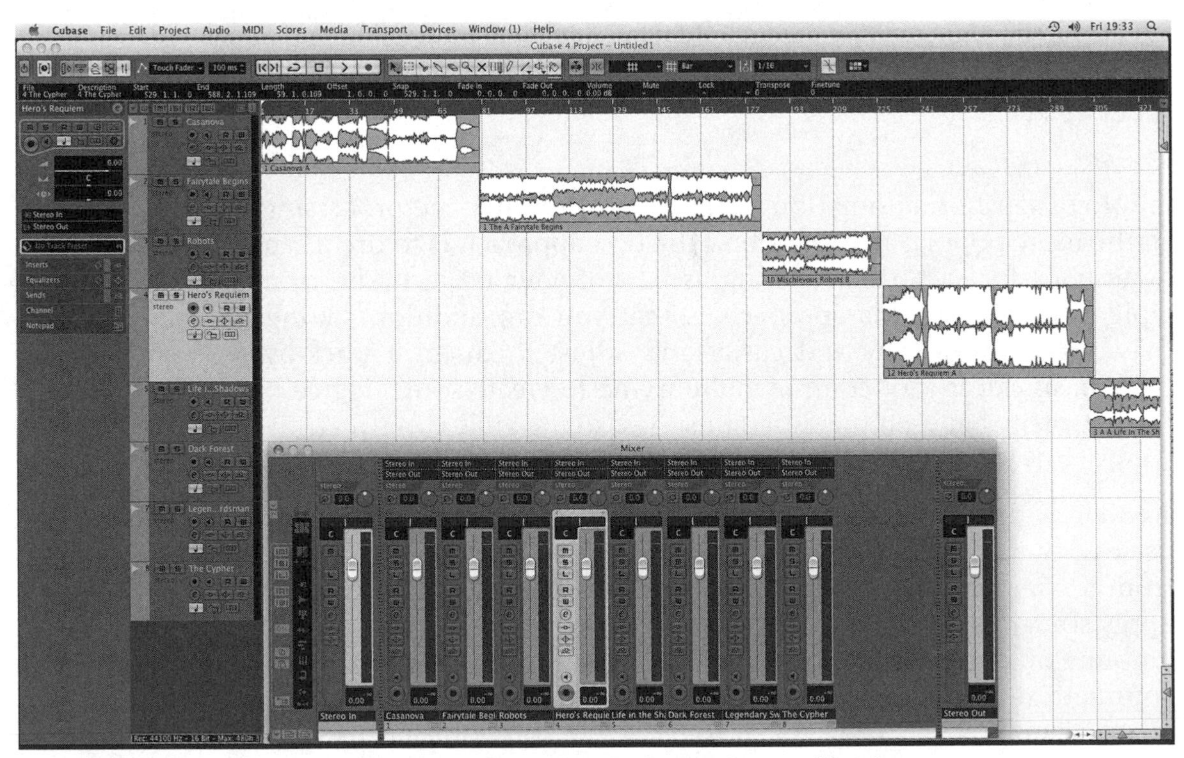

Try using a separate track for each song, allowing you to quickly audition settings between different songs.

Instantiating mastering plug-ins

Your mastering processing should now be applied on a track-by-track basis using the insert path on your DAW's channels strips. Think carefully about the order of processing – what plug-ins you decide to place at the start of the insert path, for example, and what plug-ins are placed at the end. Obviously you can also use your DAW's auxiliary send system if you need to apply effects such as parallel compression, although you need to take care if multiple tracks end up using the same auxiliary send.

Having applied plug-ins on individual channels, it's also interesting to make some distinction on the use of the stereo buss. The stereo buss is arguably the best place to instantiate any metering plug-ins that you might use, whether it's for the purposes of basic level assessment, frequency analysis, or loudness metering. Unless you're really pushed for CPU usage, you'll want to avoid placing any plug-ins here that actually process the audio.

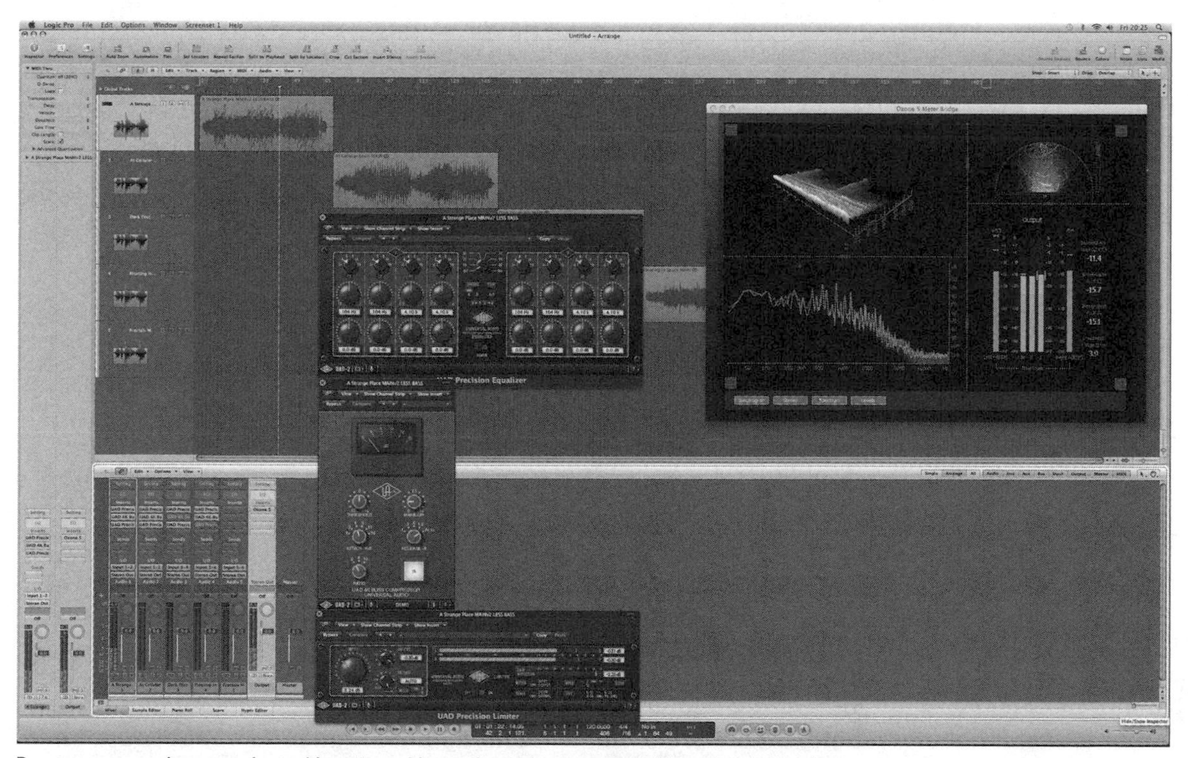

Process songs using your channel inserts and leave the main stereo buss for metering plug-ins.

Rendering pre-master files and applying fades

Once you've finished applying the processing, you then need to render your pre-master files ahead of any fade-outs or fade-ins. At this stage, you'll probably still want to keep your master in a high-resolution format, rather than applying bit reduction or sample rate modification for example. Once you've rendered the pre-mastered files these can then be re-imported into your session for you to apply any required fade-ins or fade-outs, and if appropriate, reappraise the exact start and end points for the songs.

When it comes to rendering fades and adjusting the start and end points, you'll often have two principle options – either to directly modify the audio files using some form of sample editor (in other words, you irrevocably modify the date in the original file), or you apply fades and edits using the DAW's conventional region editing tool and then form a subsequent re-bounce. Either the way, the key point is that the

Fades should be applied after the application of mastering processing, using your DAW's fade tool.

fade has been applied post-processing, so that fade doesn't interfere with or influence the application of compression and limiting.

Final delivery

For the purposes of this exercise, we're going assume you're creating final masters as 16 bit 'production-ready' audio files, rather than delivering a DDP file ready for replication (if you do need to deliver a DDP, see Chapter 9). The final stage of the mastering process, therefore, is to export your finished region in a format appropriate for delivery and/or the next stage of post production, which in most cases with be a 16 bit, 44.1 kHz .wav file. Given that you've been previously working at 24 bit, and will have applied the final fade-outs, you also need to factor in the application of dither. In most cases, you DAW will support the application of dithering as part of the export process, particularly when moving from 24 bit to 16 bit resolution.

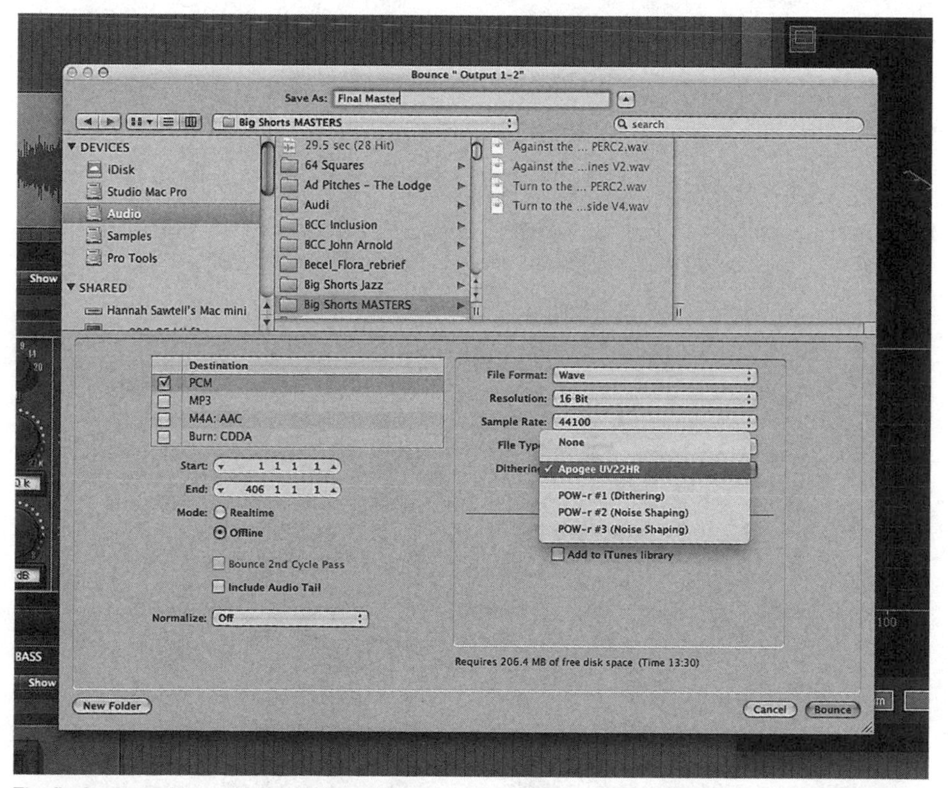

The final export will need to be at CD resolution, applying dithering to move between 24-bit and 16-bit resolution.

10.3 Method 2: Stereo buss mastering in an unconventional DAW

Not all music applications work using the horizontal track lane and channel mixer interface described, so it's important to differentiate some strategies in respect to 'unconventional' DAWs. In the absence of a full and complete mastering workflow, therefore, it might be a case of making a few well-informed compromises and at least getting proportionately closer to a 'mastering grade' output.

In situations where you don't have access to a typical mastering workflow, the main objective is to use signal processing across

In Reason, mastering processing should be instantiated after the main output of its console.

the main stereo buss so as to achieve some of the principle effects and treatments that a mastering engineer might use. In the case of Reason, for example, users have access to a number of MClass signal processors, which can be inserted after the main stereo outputs of Reason's remix console, with the eventual results being fed to the main audio outputs. Once the processing is applied, the mix is exported, ideally picking the appropriate format for the next stage of post-production or delivery.

To better understand what you can achieve in Reason, let's look at some of the principle MClass tools:

MClass Equalizer

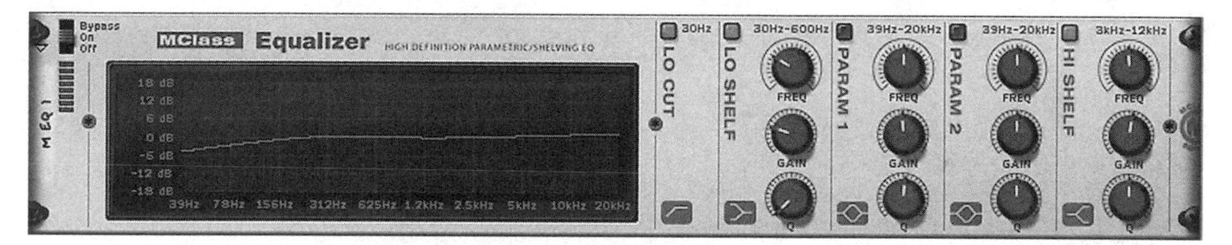

The MClass Equalizer is a simple four-band equalizer, with two shelving sections at either end of the audio spectrum, two parametric mids and a simple low-cut switch at 30 Hz. Used with care, it's a good way of bringing the timbral qualities of a Reason mix into line – whether you're notching out some low-mid weight, for example, or adding a touch more 'air'.

MClass Compressor

The MClass Compressor is a simple broadband compressor, so it's great for many of the compression techniques we talked about in Chapter 4, whether you're using an ultra-low ratio and the adaptive release to 'glue' your mix together, or using a hard ratio more focussed on to controlling the transients. Some clever routing in respect to the MClass Compressor's side-chain input and the MClass

Stereo Imager (as we'll see in a minute) can significantly extend its practical application in mastering. Newer versions of Reason, and of course, Record, also includes Propellerhead's version of the infamous SSL buss compressor (also covered in Chapter 4), which isn't permanently wired across the stereo buss of the mixer.

MClass Stereo Imager

The MClass Stereo Imager is the 'secret weapon' in Reason's collection of mastering tools, and it's an important way of extending the

functionality of the MClass Compressor. In effect, the Stereo Imager is a form of multiband M/S processor, allowing independent control over the relative width of two frequency bands – high and low. As we saw in our previous explorations of M/S, there's a lot of value to be had in converting the lowest portion of the audio spectrum to mono, as well as widening the high-end, so the MClass Stereo Imager is a great tool for creating a punchy and impressive master.

Intriguingly, the processor also features independent solo controls for both the high and low band sections, effectively allowing you to combine the MClass Stereo Imager with the MClass Compressor to create a form of two-band multiband compressor.

MClass Maximizer

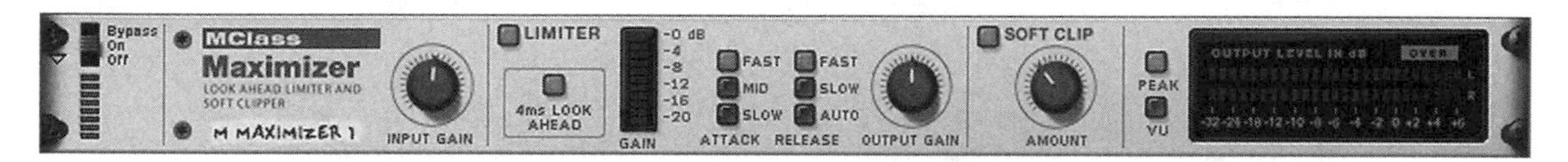

The MClass Maximizer is Reason's take on a digital brick-wall limiter, and as such should be placed as the last device in your mastering signal path. Use a fast attack and the *auto release* setting to provide the most transparent and surgically effective limiting. Use the *soft clip* to create an edgier form of limiting, where the transients are deliberately clipping rather than compressed. This 'clipped' limiting works well on electronic music and has much the same effect as a mastering engineer deliberately clipping a pair of A/D converters.

A Word of Caution

As with all buss processing, care needs to be taken not to confuse the objectives of mixing and mastering. Be wary, therefore, of applying too much 'mastering processing' while you're mixing, ideally leaving the buss as clean as possible as you piece together the mix. All too often, buss processing is instantiated to rectify deficiencies in the mix, often forming lazy solutions to a problem that can be solved in a more considered and effective way. For example, if there isn't enough energy to the mix don't just opt for a dose of buss compression, but look carefully at how the various instruments interact with each other.

Ideally, apply your buss processing at a completely different time, so that you gain some clarity from not having listened to the mix for a few hours! Make careful comparisons to other work – whether commercial CDs, or other tracks being presented at the same time – ensuring that the timbre and dynamics are consistent across the tracks. Of course, deficiencies in the mix might become apparent, so there's no harm in tweaking a few channel faders, but try to make the two activities distinct and separate from one another, as you would if your where re-importing a track into a DAW for a more conventional mastering workflow.

CHAPTER 11

Conclusions

In this chapter

11.1 The art of mastering 237
11.2 Technology as a conduit 237
11.3 Head and heart 239
11.4 An evolving art form 239

11.1 The art of mastering

Having explored a myriad of techniques and principles across this book, we finally come to the point where we need to consider the overarching skill and art of mastering. As with many aspects of craftsmanship, it's easy to get absorbed by the minutia and detail of mastering – a particular compressor setting you might use, for example, or a precise dithering algorithm – at the expense of understanding the whole. Ultimately, the art of mastering floats somewhere above all the principles and technique we've explored – when one understands that an objective can be achieved in a multitude of ways and that the music is far more important than any individual process you carry out. Put simply, mastering is more about an approach to listening than a specific plug-in or particular workflow.

11.2 Technology as a conduit

As a guiding ethos to all your work, it's worth remembering that the technology of recording ultimately acts as a conduit – the connection between the musician and the listener, rather than a complete entity in its own right. At its best, a good recording simply captures a performance or musical idea as clearly as possible, with the technology allowing the listener to enjoy this repeatedly, rather than having to attend a live performance whenever they want to listen to a piece of music! Whenever the conduit creates a barrier to the listener's enjoyment, or it becomes too noticeable in its own right, the recording (and/or its mastering) has failed to some degree.

The enjoyment of music is a guiding light that should always inform the process of mastering – understand the key emotional message contained within a track and ensure that the technology delivers this point as effectively as possible. Questions as to whether a particularly EQ setting is wrong or right, therefore, or whether a compressor is overcooking the amount of gain reduction being applied, can be answered by critically examining your decision-making process and seeing if it positively contributes to the message behind the music. In short, the moment your decisions interfere and detract from the music, you know you're in trouble!

Of course, thanks to Metallica's Death Magnetic incident we all now have a suitable reference point regarding a 'flawed conduit' – the point where aggressive mastering, or at least some ill-informed production decisions, leads to a compromised listening experience. Although there are plenty of other audio atrocities released on an almost daily basis, the Death Magnetic example perfectly illustrates the threshold where the imprint of mastering becomes too overbearing for the music it's trying to convey. Restraint, empathy and musicality are just as important to mastering as the ability to push a brick-wall limiter to the point of distortion!

As a reassuring point, it's also worth noting how good music can often transcend technical limitations or imperfections. Early jazz recordings that were cut direct-to-disc are infused with plenty of energy and musicality, often being just as enjoyable as recordings produced in more 'audiophile' surroundings. Despite the severe limitations embedded in the technology, the engineers often made simple and effective decisions that ensured the music was conveyed as directly as possible. By contrast, the fact that we have a bewildering amount of technical options – particularly in relation to the amount of choice for signal processing – doesn't necessarily mean it's any easier to arrive at an appropriate decision!

11.3 Head and heart

To a large extent, mastering needs to straddle the intersection between art and technology carefully – balancing technical considerations (such as Internet data compression, flaws of playback systems, delivery formats and so on) against musical objectives. Listening as you're mastering, therefore, involves using both your head and your heart, making both an emotional and intellectual response to what you're hearing. While this might sound easy, the ability to switch between two radically different ways of hearing can be a difficult skill to acquire, and something that may take years to master.

If there's one overarching principle of this book, it's the importance of listening, learning and adapting. Simply following the guidance in the book 'blindly' won't produce effective masters of your music (or the music that's presented to you). Instead, you'll need to listen, adjust the parameters and carefully analyze the results so that you can adapt the principles to best fit the music you're trying to master. Remember, each track presents its own unique set of issues, so a 'one-size-fits-all' solution rarely delivers appropriate results. Most importantly, though, you'll start to develop your own identity, finding combinations of plug-ins or particular sonic techniques that deliver the sound that you feel is most appropriate to recorded music.

11.4 An evolving art form

Like all art forms, music and the art of mastering is a constantly evolving entity. Mastering practices today will undoubtedly inform future techniques, but it's is also clear that music will be in a radically different place in 20 years' time – both in a stylistic sense and in the way we consume it. Indeed, few people working in the music industry predicted the radical shifts ushered in by the iTunes and YouTube generation, and as the pace of technological evolution becomes even faster few would attempt to second guess how the market-place will change. What is clear though, is that the role of mastering will need to adapt – both in respect to new sounds created by the next generation of musical innovators, and in respect to new delivery systems and standards that have yet to be developed.

Index

A

A/D & D/A converters 13, 15, 35
AAC 42, 218–20
Ableton Live 20
absorbers 34
acoustic treatment 34
AlterMedia 224
amplitude 137
analogue peak limiting 152–3
Aphex Aural Exciter 130–1
Aspera 223
assessment 48
  dynamics 49
  frequency 48–9
  perceived quality 50–1
  stereo width 49–50
ATC 34
attack and release 60–2
audio importing 226
Audio Interchange File Format 42
audio spectrum 101–2
  2-6kHz-Hi mids (bite, definition, beginning of treble) 106–7
  7-12 kHz (treble) 107–8
  10-60 Hz (subsonic region) 102–3
  12-20 kHz (air) 108–9
  60-150 Hz (root-notes of bass) 103–4
  200-500 Hz (low-mids) 104–5
  500Hz-1 kHz-mids (tone) 105–6
auto settings 62–3

B

balance
  perfect 92–3
  yin/yang of EQ 95–6
barcodes (UPC/EAN) 208–9
bass
  attenuate in side channel 159
  control of 149
Bento 224
brick-wall limiter 159
  amount of limiting one can add 159–60
broadband compression 79
buss processing 232–6

C

capture 53
catalog numbers 202, 209
CD-R 220
CD-TEXT 205, 211–12
CDA 216–17
CDs 8–9, 197, 216–17, 223–4
codes 202
colouration tools 125
  converters 128
  extreme colour 128–30
  non-linearity 125–7
  phase shifts 127
  sound of components 127
compression
  attack/release 60–2
  auto settings 62–3
  basics 59–64
  classical parallel 74–6
  controlling frequency balance 153
  gain makeup 64
  gentle mastering 69
  glue 74
  heavy 71–2
  knee 63–4
  multiple stages 76–8
  New York parallel 76
  over-easy 70–1
  peak slicing 72–3
  ratio 59–60
  threshold 59
compressors 27
  broadband 16
  FET 67–8
  gain reduction 59–60
  limiter 16
  multiband 16
  optical 65–6
  types 65–8
  variable-MU 66–7
  VCA 68
confidence 47–8
consoles 17–18
  input/output level controls 18
  monitoring sources 19
  series of input sources 18
  signal processing 18–19
  switchable insert processing 19
conversion 13, 15, 35, 42–3
converters 128
Count 210–11
Credit Is Due campaign 210–11
Credit Where Credit Is Due campaign **(MPG)** 210
cross-platform solutions 24
crossover 17
Cubase 20, 226
cut filters 18–19
cutting engineer 5

D

data compressed audio 219
Daylite 224
DDP Creator (Sonoris) 205
DDPi (Disc Description Protocol image) 56
de-essing 120
delivery 55–6, 220–3
delivery file formats 214–15
depth 178–9
DigiDelivery format 221, 223
Digital Audio Tape (DAT) 212, 214
digital audio workstation (DAW) 3, 12–13
  off-the-shelf 20–1
  stereo buss mastering in unconventional DAW 232–5
  track-based mastering in conventional DAW 226–31
digital over 161–2
Disc Description Protocol (DDP) 13, 22, 205, 215–16, 221
distortion 163–4
dither 42–3, 46, 52
Dorrough meter 25

DVD-R 56, 220
dynamic range 135–6
dynamics 54–5
  assessment of 49
  basics of compression 59–64
  compression in M/S dimension 85–6
  compression techniques 69–78
  controlling 58–90, 149–50
  expansion 86–90
  from broadband to multiband 79–80
  moderating overall song dynamic 151
  setting up multiband compressor 80–5
  shape them, don't kill them 148
  types of compressor 65–8
  using side-chain filtering 78–9

E

editing 181–2
  fades on tracks 187–9
  gaps between tracks/sonic memory 184–7
  hidden tracks 199
  level automation 191–3
  markers 196, 198
  sequencing 182–3
  seques 190–1
  snapshots 193–4
  stem mastering 194–6
  topping and tailing 183
  track IDs 196, 198
  types of fades 189–90
encoding 209–11
enhancers 130–1
equalization (EQ) 92–3
  combined boost/attenuation 114–15
  controlled mids using parametric EQ 115–17
  cut narrow, boost wide 117–18
  fundamental v. second harmonic 118–19
  high-pass filtering 109–10
  loudness 154–8
  mid-channel 124–5
  removing sibilance 119–21
  selective 121–2
  shelving 111–12
  side channel 122–4
  smiling EQ 155
  strategies 109–21
  understanding curve of 112–14
  yin/yang of 95–6
equalizers 20, 27
  dynamic 120–1
  filtering 99
  graphic 99–100
  non-symmetrical 100–1
  parametric 98
  phase-linear 100
  shelving 97
  types 97–101
excitement 163–4
  theory 143, 145
exciters 130–1
expansion 86
  downwards 86–9
  upwards 89–90

F

fades 187–9
  applying 230–1
  types of 189–90
Fairchild 27
Fast Fourier Transform (FFT) 32, 118
file formats 42–3
File Transfer Protocol (FTP) 221
Filemaker database 226
filtering 99
  high-pass 109–10
  side-chain 78–9
flow 39–40
frequency
  balance of instruments 94
  controlling balance with multiband compression 153
  decoding problems 94–5
  loudness and 140–3
  low, mid, high 94
  technical problems 95
  unwanted resonances 94–5

G

gain makeup 64
gain reduction 59–60
gap time 184–7
Gracenote/CDDB 211, 224

H

harmonic balancing 156–8
harmonic structure 118–19
hidden tracks 199
high-mid frequencies (HMF) 98
Home Taping is Killing Music campaign 213
human hearing range 109

I

index points 197
inflation 163–4
inter-sample peaks 164
International Standard Recording Codes (ISRC) 13, 202–5
Internet 2, 38, 56, 210, 213, 214, 216, 219, 220, 239
iTunes 42, 48, 213, 222–3
iZotope Ozone (TC Electronics) 24–5, 26, 52

K

knee 63–4

L

law of diminishing returns 145–7
level automation 191–3
listening 41
  holistic, micro, macro foci 44–5
  interface/quality benchmark 41–2
  subjectivity vs objectivity 43
  switches 45–6
Little Lab's Voice of God 114–15
logarithmic scale 189–90
Logic 20
Logic Pro 21–2
loudness 7, 16, 134–5
  attenuate bass in side channel 159
  brick-wall limiter 159–63
  classic mistakes 150–1
  compress mid channel 159
  control bass 149
  control dynamics before limiting them 149–50
  controlling excessive mid 156
  controlling frequency balance with multiband compression 153
  duration/transients 138–40
  dynamic range in real world 135–6
  equalizing for 154–8
  excitement theory 143, 144
  frequency and 140–3

harmonic balancing 156–8
inflation, excitement, distortion 163–4
law of diminishing returns 145–7
metering 144–5
moderating overall song dynamic 151
moderating transients with analogue peak limiting 152–3
multiband limiting 163
over-limited chorus 150–1
power of mono 148–9
principles of perception 136–7
rolling off the sub 154–5
secret tools 163–7
setting output levels 162–3
shape the dynamics 148
simple/strong productions 147–8
smiling EQ 155
sound checking 152
stereo width and 158
low-mid frequencies (LMF) 98

M

magnetic tape 6
Manley 27
markers 196, 198
mastering 19
art of 237
CDs 8–9
concepts 1–2, 38–9
creative 8
cutting/need for control 6–8
delivery 55–6
early days 5–6
effective 38
essential improvements 33–5
evolving art form 239
final checks 223–4
head and heart 239
history/development 5
key skills 3, 5
personnel 39
preparation 202
process 51–3
role 39
step-by-step guide 4–5
strategic levels 39
technology as conduit 237–8
understanding 2–3
MClass Compressor 233–4
MClass Equalizer 233
MClass Maximizer 235
MClass Stereo Imager 234–5
MD5 216
Metallica, Death Magnetic 238
metering, peak vs RMS 25
mid/side (M/S) 18–19
attenuate bass in side channel 159
compress mid channel 159
compression 85–6
dedicated plug-ins 31
manipulation for width 175–6
mid signal 86
processing 29–30
selective equalization 121–2
sides of the mix 86
stereo plug-ins 30–1
tools 30–1
mix 167
monitoring 16–17, 34–5
mid-field/far-field 16–17
near-field 17
sources 19
mono 148–9
MP3 42, 211, 213, 218–20
multiband compression 79–80
behaviour across the bands 81–2
controlling bass 82
controlling frequency balance 153
controlling highs 83–4
controlling mids 84
crossover points/amount of bands 80–1
extreme colour 128–30
gain makeup 84–5
setting up 80–5
multiband limiting 163
Music Producers Guild (UK) 210

N

NARAS 210
non-linearity 125–7

O

output levels 162–3
over-limited chorus 150–1

P

PCM (pulse code modulation) audio 42
phase 172, 174
phase meter 32
phase shifts 127
planning 48
PlexTools 217
plug-ins 24
instantiating mastering plug-ins 229
M/S 30–1
PMC 34
PMCD (Pre Master CD) 217
POW-r suite of algorithms 52
PQ sheet 205–8
pre-emphasis 212
pre-mastering 9, 230–1
Pro Tools 20, 226
processing 53–4
Pseudo Stereo 177
Pultec EQP-1 A Equalizer 114
Pyramix 12, 21, 187, 190, 198, 204

Q

quadrophonic (quad) 171
quality assessment 50–1
quality control 210

R

ratio 59–60
Reason 20
Recoil Stabilizer (Primacoustic) 34
recording engineer 5
Red Book 13, 21, 192, 202, 205, 216, 220, 225
reel-to-reel recorders (Studer) 27
reflections 34
reverb 179–80
RIAA 213–14

S

Sadie 12, 21, 187
sample rate conversion (SRC) 42–3, 55
sequencing 54–5, 182–3
seques 190–1
Sequoia 187
Serial Copy Management System (SCMS) 212–14
shape, shaping 40–1, 167
sibilance 119–21
side-chain filtering 78–9
signal processing 18–19
Slate Digital Virtual Tape Machines 166–7
snapshots 193–4
software 20–8

songs to tracks 226–8
sonic memory 184–5, 187
Sonic Solutions (SoundBlade) 12, 22, 187, 217
Sonnox Inflator 165
Sonoris 205
Sony 1610 (later 1630) converters 9
sound checking 152
Sound Forge Pro 10 (Sony) 24
sound pressure levels (SPL) 137
soundBlade HD 22
soundBlade LE 22, 23
soundBlade SE 22
space 169–80
spectral, frequency-based analysis 32
stem mastering 194–6
stereo
  adjustment 18–19
  delay 176–7
  metering 172–3
  vectorscope 32
stereo buss mastering 232–3
  MClass Compressor 233–4
  MClass Equalizer 233
  MClass Stereo Imager 234–5
  MClass Stereo Maximizer 235
Stereo Controller 178
Stereo Enhancer 177–8
stereo width 49–50, 158–9, 172
  formats/current practices 171–2
Stereopack (PSP) 177–8
StudioSuite (AlterMedia) 224
surround sound 171
switchable insert processing 19

T

T-RackS 3 (IK) 27–8
threshold 59
timbre 92–3
  audio spectrum 101–9
  balance 95–6
  colouration tools 125–8
  decoding frequency problems 94–5
  exciters/enhancers 130–1
  multiband compressor 128–30
  selective equalization 121–5
  strategies for equalization 109–21
  types of equalizer 97–101
topping and tailing 183
track IDs 196, 198
track names 202
track-based mastering 226
  final delivery 231
  importing audio 226
  instantiating mastering plug-ins 229
  rendering pre-master files/applying fades 230–1
  songs to tracks 226–8
transients 139–40
  moderating with analogue peak limiting 152–3

U

U-Matic tape 9
Universal Audio Precision Maximizer 167
Universal Audio UAD system 26–7

V

VHS 197
vinyl cutting 6–8

W

WaveBurner (Apple) 21–2
Wavelab (Steinberg) 24
Waves L-316 28
Waves Maxx Bass 131
WeTransfer.com 221
width 169–70, 170, 176–8
word length reduction (WLR) 52

Y

YouSendIt.com 221